The Racism of Cybersecurity

A Solution

by

William Benjamin Jones

The Racism of Cybersecurity

info@networklawandorder.com

ISBN: 979-8-234-01397-2

CONTENTS

To: Pola

PART 1

CRIME SCENE DO NOT READ

THE CRIME

Chapters 1-6

CHAPTER 1

THE CONCESSION SPEECH

November 9, 2006, Carlyle House, Alexandria, Virginia

I stood on the manicured grounds of the historic Carlyle House in Old Town Alexandria, iced tea sweating in my right hand, watching the media gather and circle. An hour earlier, the call had come: Senator George Allen was finally conceding. The speech was scheduled for 3:00 p.m.

It was my wife's day off. "Come with me," I'd said. She'd looked at me with that expression — the one that said she knew this mattered but didn't quite understand why. We found parking adjacent to the Carlyle House. Small miracle. What a waste, I thought. What more could I have done?

The question hung in the November air like the cigarette smoke from the nervous staffers clustered by the entrance. Print journalists checked their watches. Television crews adjusted their cameras. Everyone knew what was coming. Senator George Allen — former Governor, current U.S. Senator, presumptive 2008 Republican presidential frontrunner — was about to politically die. And I was watching three other casualties that no one could see.

I scanned the crowd. A sea of White faces punctuated by a handful of Black technicians operating cameras. Was I a traitor to my people? Did the press think I was a confused Republican wannabe? Did anyone understand that I was standing here with my fellow victims?

Six hundred and fifty thousand youngsters could have been employed in high-tech jobs that couldn't be outsourced. Just like the Internet, it would have all started in Virginia. New jobs. American jobs. Jobs for kids aged 16-22 who'd learn to protect our critical infrastructure while building careers that would last a lifetime. What a waste.

The doors to the Carlyle House remained closed; it was 3:14 p.m. now. The ice in my tea was melting. My wife stood on the other side of the gathering, slightly out of place, patient as always. She'd heard me talk about Network Configuration Management (NCM) until the words probably haunted her dreams. She knew about the meetings. The briefings. The years of "Yes, we're very interested" followed by silence.

She didn't know about the phone call I'd gotten two months earlier. The president of a small telecom business, tapping my shoulder at the Independent Pioneer Telephone Association conference: "Ben…." He told me he had been warned not to do business with me and that his contracts might be in jeopardy. I'm willing to sign a statement to that effect." That was when my real education began.

At last, the door opened. Senator Warner emerged first, his wife beside him. Then Senator Allen's children — restless, slightly annoyed the way teenagers get when forced into adult ceremonial duty. Finally, Senator and Mrs. Allen descended the stairs to take their places on stage. The adults looked weary. Defeated.

Senator Warner took center stage to introduce his fallen colleague. His voice was steady but sullen, praising Allen's contributions to the Commonwealth, their work together, and Allen's technology initiatives. Technology initiatives. I almost screamed. The irony was suffocating.

Two weeks earlier, I'd attended a campaign rally for the Republican ticket at a Springfield trucking company parking lot. The crowd had been energized. More Black faces than at today's subdued gathering.

Senator Allen had thrown his signature football into the crowd — the ritual he performed at every event. Today was no different. He looked in my direction and threw the football right at me. Iced tea in my right hand, I caught it with my outstretched left. The crowd applauded. The Senator raised his eyebrows in approval. Look at me, catching passes from a dying political career.

Now Senator Warner was wrapping up his introduction, his praise of Allen's "technology initiatives" hanging in the air like a bad joke. I took one last look around. Where was Congressman Tom Davis? Another victim. If Senator Warner, Senator Allen, Congressman Davis, and I were victims, there had to be a driving force behind it. The applause was polite but muted. Senator Warner hugged Senator Allen and stepped aside.

Senator Allen approached the podium, football still in hand. His concession speech was themed "A time for all seasons." I watched his face — the despair, the confusion. He didn't know what had happened to him. Neither did Warner nor Davis. They were victims and didn't even know it.

The speech ended with sympathetic, polite applause. A staffer invited campaign workers and supporters to the back of the building, where the Senator wanted to give personal thanks.

Before I could reach my wife, the press contingent swarmed me. "Is Senator Allen a racist?" "Why did you support someone with racial baggage?" "Did the Macaca thing bring down the Senator?" "Can you explain why so few Blacks supported Republicans in this election?" Five reporters deep, all holding microphones and recorders, all hunting for sound bites.

I kept my composure. Measured responses. No, I didn't think

the senator was racist. Yes, a few Blacks supported the Party, but the Republican message to minorities hadn't been framed correctly. But how could I tell them what I really wanted to say?

How could I explain that these congressional staffers had lost power? That I'd lost money? That the country had lost a way to galvanize disenfranchised minorities, create jobs that couldn't be outsourced, and mobilize youth to protect critical infrastructure?

How could I tell them about the meetings — not one, not five, but dozens over six years? How could I explain sitting next to Senator Warner in his office, hearing him say, "Ben, forget about the jobs in Northern Virginia. Let's just get you the contract first. You've been around here for years, and fellows," looking at his aides against the wall, "let's get this thing done." Except it never got done.

How could I tell them about Donald Rumsfeld's chief engineer leaning forward in that Crystal City meeting and saying, "Ben, if you can really do this, hell, the Pentagon will erect a statue of you in their front yard."

How could I explain that three-star generals — not one, not two, but three — had addressed/validated the claimed benefits of NLO marketed it, even selected it for their commands, only to be overruled? The press was looking for sound bites about Macaca.

We were victims of something far more sophisticated, far more insidious, and no one cared. Senator Allen had lost by 2,870 votes. Less than three thousand votes in a state with millions of voters. Less than one-tenth of one percent.

If these legislators' staffers had understood the significance of what we'd presented to them on no fewer than ten occasions dating back to 2001, no amount of Macaca references could have jeopardized the outcome. People vote with their pocketbooks.

Six hundred and fifty thousand new jobs would have swung Virginia by a landslide. I answered a few more questions, said nothing that mattered, and worked my way through the crowd to find my wife. The reporters moved on to more promising targets.

As we walked back to the car, she squeezed my hand. "Are you okay?" "Yeah," I lied. "I'm okay." But I wasn't okay. None of us was okay.

Senator Allen had just lost his seat and any hope of the presidency. Senator Warner would soon lose his chairmanship of the Armed Services Committee. Congressman Davis would lose his chairmanship of the House Government Operations Committee. The Republicans would lose control of Congress. And the country had lost something it didn't even know existed.

The question that would haunt me for years was simple: How did this happen? The answer was even simpler: T.C.M. Trivialized. Compromised. Marginalized. And the institutional culture that did it to us — to all of us — was about to get away with it. Unless someone told the story.

Chapter 1 reflects the author's personal recollections, impressions, and opinions regarding events he directly experienced. Interpretations of motives and outcomes are presented as personal perspective rather than assertions of fact.

CHAPTER 2

NETWORK LAW & ORDER®
THE NEW OIL

"Digital Gold at the Ground Floor: It's Not Like Oil... It Is Oil."

In 1870, John D. Rockefeller founded Standard Oil. By 1880, Standard Oil controlled 90% of America's oil refineries. Rockefeller became the richest man in history not because he found oil — others had done that — but because he understood that crude oil was useless without the refining infrastructure to make it valuable.

He built the infrastructure. He controlled the standard. He captured the value.

Standing at the Carlyle House in November 2006, watching 650,000 jobs evaporate with Senator Allen's concession speech, I realized we're at the exact same moment Rockefeller faced in 1870.

The crude oil exists. It's everywhere. It's in every network, every database, every connected system in America and around the world. But it's useless — worse than useless, it's dangerous — without the refining infrastructure to make it valuable. JGI LLC invented that infrastructure. We hold the intellectual property.

The Market Is Larger Than Petroleum

Every network on earth could materially benefit from what we've invented.

Every military network.

Every telecommunications carrier.

Every financial institution.

Every hospital system.

Every government agency.

Every corporation with IT infrastructure.

Every cloud provider.

Every data center.

The global network security market alone is projected to exceed $500 billion by 2030. But that's just "security" — the ambulance at the bottom of the cliff. NCM — what I call **Network** Law & **Order** — is the fence at the top. It's the preventive medicine. It's the thing that makes all those security tools actually work.

Without NCM, organizations are spending $1.65 for every 98 cents of value. They're running engines without knowing which parts connect to which other parts. They're flying blind.

The total addressable market isn't $500 billion. It's every dollar spent on network operations globally. We're talking trillions, not billions. And right now, that market has no standard, no dominant player, no Rockefeller. Until now.

Why the last decade has been a Ground Floor Moment

A lubricant is "a substance, typically oil or grease, applied to a surface to reduce friction between moving parts." The **Network Law & Order** business process is digital oil that reduces the friction between the design network parameters and the operating network parameters — but it's physics-based, not petroleum-based. Just like Rockefeller's moment, three forces converge to create inevitable demand:

First: The Technology Storm Has Arrived

In the 1980s, three elements married computers and telecommunications:

Legal: 1984 AT&T deregulation spawned new competition.

Scientific: Shift from Silicon to Gallium Arsenide (1000x speed increase).

Engineering: Very Large Scale Integration (VLSI), putting entire systems on chips.

This created Net-Centricity — the requirement for multiple networks to work together seamlessly. Think of the TV show *24* where CTU's control room displays real-time data from motor vehicle databases, hospitals, telephone networks, insurance companies, military systems, and police networks, all working together.

That's not science fiction anymore. It's the minimum requirement for modern defense, healthcare, finance, and commerce. But here's the problem: You can't achieve Net-Centricity without Network Oil. Without NCM, networks overheat from friction. Parts crack. Attackers exploit the cracks. The whole system fails.

Second: Foreign Competition Is Forcing the Issue

The United Arab Emirates' Etisalat said it explicitly in its 2004 annual report:

> The basics of reliability, consistency, and availability continued to be the main features of Etisalat's core fixed line network. Those qualities combined to play an important role in encouraging the influx of global and regional businesses, making the UAE the base for their Middle East operations and, in many cases, the location for their headquarters offices.

Reliability. Consistency. Availability. Those are the three viscosity properties (ingredients) of Network Oil. The UAE understands this. Japan understands this. South Korea understands this.

They are building telecommunications advantages on the foundation of NCM while the United States has moved cautiously.

When foreign carriers offer demonstrably more reliable networks than American carriers, global businesses relocate their headquarters accordingly. We're not just losing market share — we're losing the commanding heights of the digital economy.

Third: National Security Demands It

After 9/11, the Pentagon couldn't efficiently rebuild its telecommunications infrastructure because it didn't have a baseline. They didn't know what they had before the attack. No configuration records. No Network Oil.

The defense establishment validated **Network** Law & **Order** in a panel of government technology experts' review on April 19, 2001. Two former directors of DISA endorsed it. The Chairman of the Joint Chiefs of Staff ordered it briefed to his three-star communicator, Lt. General Joseph Keith Kellogg, now President Trump's incoming Advisor to Ukraine, who was in those briefings.

A principal reason NLO isn't deployed across DoD right now is that the military-industrial complex did not implement a solution they didn't control. They'd rather spend $1.65 for every 98 cents of value than let an outsider capture the market.

Why Digital Oil Is MORE Pervasive Than Petroleum

Here's what makes Digital Oil fundamentally different from

crude oil — and why the market opportunity is actually larger than Rockefeller's:

Not everyone owns a car. Not everyone uses petroleum products daily. But everyone uses networks. Every person with a smartphone, every student in a classroom, every patient in a hospital, every transaction at a bank, every citizen interacting with government all of it runs on networks.

You can opt out of driving. You cannot opt out of the digital economy and remain a functional member of modern society.

This complete penetration into the social fabric creates something unprecedented: the need to educate an entire generation in the maintenance and operation of Digital Oil infrastructure.

Rockefeller didn't need to train a generation of oil workers — petroleum extraction and refining were industrial work done by specialists. But Network Oil touches every aspect of daily life, generates complexity at scales no previous infrastructure has reached, and requires something new:

The Artificial Intelligence — Human Intelligence (AI-HI) Framework

Artificial Intelligence cannot replace Human Intelligence, nor do we want this to happen. Artificial Intelligence is revolutionizing network design. AI can analyze traffic patterns, predict failures, optimize routing, and generate network configurations with speed and sophistication no human could match. It is the perfect engine for establishing Network Law. But here's what AI cannot do: AI cannot implement Network Order.

Networks aren't just designed, they must be operated, maintained, verified, and defended at all times. That requires human intelligence working in concert with artificial intelligence.

AI produces the network DESIGN (the "Law" — the intended configuration). HI implements the network DESIGN (the "Order" — the physical configuration, change management, verification, defense). Clean networks require the order to adhere to the law.

This isn't a job that gets automated away. This is a job that becomes MORE critical as AI advances. The more sophisticated AI-generated network designs become, the more essential human oversight becomes. Someone must verify the AI's recommendations. Someone has to catch the edge cases. Someone has to recognize when the network behavior deviates from the intended design. Someone must defend against attacks that exploit the gaps between design and implementation. That someone is a Cyber Cleaner trained in NCM.

The AI Boom Makes NCM Essential

Every data center supporting AI requires massive network infrastructure. ChatGPT, autonomous vehicles, smart cities, IoT devices — all of it runs on networks whose complexity is growing exponentially. The AI revolution doesn't reduce the need for Network Oil. It multiplies it.

Every AI model training run generates petabytes of network

traffic. Every edge computing deployment requires configuration management. Every autonomous system requires network reliability that approaches 100% uptime.

Without NCM, AI infrastructure collapses under its own complexity. Increasingly powerful engines are being run without knowing which parts connect to which other parts. The friction increases. The heat builds. The cracks appear, and attackers exploit those cracks.

Case In Point: Maryland

This Creates an Unprecedented Workforce Opportunity

The 375,000 young people in Maryland aged 16-22 who play video games aren't just potential workers. They're the ONLY workforce that can scale fast enough to meet AI-era demand for network operations.

Video gaming has already trained them in:

Spatial reasoning (understanding network topology)

Pattern recognition (identifying anomalies)

Rapid decision-making under pressure (incident response)

Collaborative problem-solving (team-based operations)

Sustained attention to complex systems (monitoring)

These are precisely the skills NCM requires. NLO is not asking them to learn something alien to their experience. We're asking them to apply skills they've already developed to infrastructure that affects their daily lives.

The HI-AI framework means these aren't jobs that disappear when AI gets better. These are jobs that become MORE valuable as AI advances. The better AI gets at designing networks, the more critical human intelligence becomes at implementing and defending those networks.

The Alternative — And Why It's Catastrophic

If we don't build the HI-AI framework for Network Oil, here's what happens: AI systems design increasingly complex networks. No one can implement those designs reliably. Networks fail under their own complexity. AI infrastructure proves unreliable. American AI companies move operations to countries with better network infrastructure — countries that implemented NCM a decade earlier.

And those 375,000 Maryland young people — the ones who could have been Cyber Cleaners earning middle-class wages in jobs that grow MORE valuable as AI advances — they're instead competing for service jobs that AI actually CAN automate.

We train them for obsolescence instead of training them for the one infrastructure job that becomes more critical as technology advances.

The New Oil affects everyone. The New Oil requires a new generation. The New Oil becomes MORE essential as AI advances. The only question is whether we're smart enough to see it.

Chapter 2 reflects the author's personal recollections, interpretations, and professional opinions. Statements regarding markets, institutions, and historical parallels are presented as perspective and analysis, not as assertions of provable fact or allegations of misconduct by any individual or organization.

CHAPTER 3

THE GENERALS... and such

"Can I see you for a minute, in private?" Diane McCoy's voice was tight. Controlled. The kind of control that barely contains fury. We stepped into the hallway outside the Defense Information System Agency (DISA) conference room. I could sense she was upset. I could see Air Force Brigadier General Bernie Schoch and his staff were still discussing the **Network** Law **&** Order briefing among themselves. Cliff and his son were packing up equipment as they exchanged confused glances.

"Don't you ever, EVER embarrass me like that again." Diane's finger was pointed at my chest. "You don't ask the General for a contract like that! That's not how it works. There is a thing called acquisition and procurement!"

I stood there, genuinely confused. "But he was blown away with it. He wanted his engineers to see it. He admitted his people could not engineer the capability, and it was ready now, so why shouldn't I get paid?"

Diane was one of the most powerful Black women in defense acquisition —a Chief Acquisition Officer at DISA, protégé of the legendary Dorothy Height, president of the National Council of Negro Women. The only pink building on Pennsylvania Avenue belonged to Dorothy Height, and inspired by Dorothy Height's mentorship, Diane opened doors for me. And I had just embarrassed her.

The meeting with DISA Deputy Commander Brigadier General Bernard Schoch had gone exactly as planned. He'd watched the briefing, asked interesting questions, and expressed amazement at what NLO could do. He'd admitted — right there in front of his staff — that his people couldn't engineer this capability. Then he looked at me with what seemed like genuine curiosity and asked how we might work together, and I, like the country bumpkin I was, asked him to give me a contract.

Bottom left: Lt. Gen. Joseph Keith Kellogg—NSC Chief of Staff, Acting National Security Advisor

Bottom right: Lt. Gen. Alonzo E. Short Jr.—Director DISA (1991-1994)

Middle Left: Gen. Hugh Shelton—14th Chairman, Joint Chiefs of Staff (1997-2001)—the highest level of institutional visibility possible

Middle Right: Brigadier General Velma Richardson—NETCOM

Top Left: Gen. Lester L. Lyles—four-star, Vice Chief of Staff USAF, Howard University graduate

Top Right: Lt. Gen. Albert J. Edmonds—Director DISA (1994-1997)

General Schoch deferred to the acquisition process. "Well, I suppose we'd need to figure out the procurement aspects...." He glanced at Diane, setting up the trap perfectly.

Diane had to intervene. She had to explain that's not how military contracting works. She had to pull me aside afterward and dress me down in front of her confused White assistants, who couldn't understand why this high-powered Black woman who'd gotten us in was now furious.

General Schoch saw what the agency needed and identified the three individuals he had tasked to develop what he was now looking at. "Dead," I said to myself. Unless I could sweet-talk these guys into believing that their work would yield my invention, I was "Dead on Arrival." In fact, the trio spent the next two years ducking and dodging the NLO.

"Work with my three engineers," he'd said before we left. "They'll help you understand the requirements process." The three engineers had no interest in supporting me. None. By refusing to help, they were admitting their own failure to solve the configuration management problem — but they'd rather live with that failure than help a Black inventor succeed where they'd failed. That was 1999: my first flag officer briefing. My first lesson in how the game was really played. But it wouldn't be my last.

Chairman of the Joint Chiefs of Staff — The Buffet Line

"Cliff, I have a new best friend, and I met him IN THE BUFFET line! See, that's Shelton, he's the Chairman of the Joint Chiefs of Staff!" "Holy shit," was Cliff's response.

Cliff Shoemaker, my vice president, told me to black-tie the affair, which turned out to be the Armed Forces Retirement Home's annual reception. The Washington D.C. campus — historically known as the US Soldiers' and Airmen's Home — was hosting its annual event, and the place was packed with retired brass and active-duty officers, all lined up for prime rib and conversation. "Know what I'm thinking?" Cliff asked.

My White VP — brilliant lawyer, excellent strategist — had no idea the Chairman of the Joint Chiefs of Staff would be attending.

But I knew. I'd just spent five minutes in the buffet line next to General Hugh Shelton. I could sense he wanted to know who I was. I said I was an inventor invited by my vice president. I casually mentioned I was aware he was the first "snake eater" to accomplish that feat.

Snake Eater is a name given to Special Forces. General Shelton had come up through Special Operations, and he was the first officer from that community to become Chairman — the highest-ranking military officer in the U.S., principal military advisor to the President, the man who sat in the Situation Room when decisions about war and peace got made, and he'd been genuinely pleased that I knew his background.

The photograph from that evening shows three men in formal attire. On the left, Cliff in his tuxedo. On the right, General Shelton, in a dark suit. In the middle, I smiled, completely unaware that this casual conversation was about to change everything. "So what do you do?" Shelton had asked in the buffet line.

Standard cocktail party question. I gave him the standard answer: "I am the inventor of **Network** Law & **Order**, but I'm having trouble with DISA."

Shelton's expression changed immediately. "Huh! They don't want it unless it was invented by DISA." One sentence. The Chairman of the Joint Chiefs had just diagnosed the entire problem.

Not-Invented-Here syndrome. The tribe is protecting its own. DISA — the three-star command that runs military telecommunications — refusing to adopt solutions that didn't originate within their walls, no matter how good those solutions were, no matter how badly they needed them.

Shelton understood this immediately. He'd been around long enough to recognize bureaucratic tribalism when he saw it. He'd probably fought it his entire career in Special Forces — the snake eaters were always outsiders challenging conventional military wisdom.

"Tell you what," he said, pulling out his card. "Give me your card, and I'll have my J6er take your brief. That alright?"

I was stunned. The Chairman of the Joint Chiefs of Staff was personally ordering his J6 — the Director of Command, Control, Communications, and Computer Systems — to receive a briefing on the **Network** Law & **Order** business process. Not suggesting. Not recommending. Ordering. I gave him my card.

So returning from the buffet line, I told Cliff of the encounter as he gave me a "Yeah, right" look, and as Shelton casually strolled up to our table and said, "So this is your VP?" I made introductions and

took photos.

Cliff Shoemaker, me, and Gen. Hugh Shelton

General Lester Lyles, Commander – AirForce Materiel Command (AFMC) – Dayton, Ohio, the Morning After

The next morning, we were on our way to present NLO to the Air Force Materiel Command (AFMC) at Dayton, Ohio, while basking in the glow of meeting the Chairman of the Joint Chiefs of Staff the night before, when the call came.

"This is the office of Lieutenant General Keith Kellogg calling for Mr. Jones, President of the Jones Group International." The J6 for the Joint Chiefs of Staff was coming on the line. "When can you come in to brief me on NLO?"

That's how fast things move when a four-star Chairman orders action. No bureaucratic delay. No layers of staff coordination. The Chairman says do it, you do it.

When we arrived at Wright-Patterson Air Force Base in Dayton, Ohio, most of the CIOs (Chief Information Officers) under General Lyle's command was in the audience. This wasn't a courtesy briefing. This was General Lyles saying to his senior technology officers: "Pay attention to this." The NLO presentation was well received. The CIOs understood immediately what we were offering — systematic configuration management that would finally bring accountability to their networks.

But like the others, General Lyles never publicly endorsed NLO. He'd seen it. His people understood it. But public advocacy? That was a different matter.

Lieutenant General Joseph K Kellogg, Director, Command, Control, Communications, and Computer Systems

One week after the Dayton, Ohio, presentation, we presented to Lieutenant General Joseph Kellogg the **Network** Law & **Order** (NLO) business process. Showed him the systematic approach to NCM. Explained how NLO created baselines, tracked changes, and maintained accountability for how networks were actually configured versus how they were supposed to be configured.

Kellogg asked good questions. He understood the problem — as J6, he briefed the Chairman on the status of military networks worldwide. He knew exactly how messy configuration management was, how much money got wasted on unaccountable infrastructure, and how security vulnerabilities hid in the gaps between documentation and reality. He took the brief because the Chairman ordered him to, and then... silence.

He had to be aware of the Senate-mandated Panel of Experts briefing coming up. In fact, our spy secretary told us his office called in the middle of the Panel of Experts briefing.

Years later, Keith Kellogg would be named a special presidential envoy of President Trump's incoming National Security Team in various capacities. The general who'd briefed NLO back when the Chairman ordered it would now have the President's ear on matters of national security and critical infrastructure protection.

THE JOINT STAFF
WASHINGTON, DC

Reply ZIP Code:
20318-6000

DEC 21 2000

Mr. Ben Jones, President & CEO
The Jones Group International
13171 Morning Spring Lane
Fairfax VA 22033

Dear Mr. Jones,

I have reviewed the marketing material you sent me a week after our 29 November 2000 meeting. I have also received a copy of the letter you sent the Chairman on 18 December 2000.

I will ensure your entire package is sent immediately to the Small and Disadvantaged Business Utilization Office, which was established to assist firms like yours to sell goods and services to the Department of Defense. For you convenience, that address is provided below.

Small and Disadvantaged Business Utilization Office
Office of the Secretary of Defense
1777 North Kent St.
Arlington, Virginia 22209
Phone: 703-588-8631

I appreciate you time and effort in this matter.

Sincerely,

JOSEPH K. KELLOGG, JR.
Lieutenant General, USA
Director for Command, Control, Communications and Computer Systems

Copy to:
Director, DISA

But in 2001, he was the J6 who took the brief because he had to, and then stayed silent because career survival required it.

Lieutenant General Alonzo T. Short, former Director of DISA Crystal City, February 26, 2001

I will leave out how I was able to get General Short to act as my marketeer because the story of how Zoom and MS Teams were created would be a book in and of itself.

John Osterholz's office in Crystal City had the antiseptic feel of all Pentagon-adjacent buildings efficient, secure, utterly forgettable. But what happened in that meeting was anything but forgettable.

Osterholz was the Secretary of Defense's chief engineer. I am sure John had a fancier title than chief engineer, but when the Secretary of Defense needed technical advice, Osterholz provided it. Senator Warner's staffer had arranged the meeting, and we'd brought reinforcements: Lieutenant General Alonzo T. Short, former Director of DISA, now President of Houston and Associates. General Short wasn't just attending. He was vouching for us.

He demonstrated NLO. Showed how systematic configuration management could finally bring accountability to Defense networks. Explained the business process — not just tools, but the engineered approach that would work regardless of which tools you used.

Osterholz leaned forward. "Ben, if you can really do this, hell, the Pentagon will erect a statue of you in their front yard." We all laughed. But there was an edge of seriousness to it.

General Short didn't laugh. He looked directly at Osterholz and said, "John, my best engineer has looked at this technology, and he, like I, is convinced we have something here." Not "interested." Not "promising." Convinced. And "we" — as in, he was personally invested in getting this deployed.

Sixteen months later, on June 12, 2002, at 11:23 PM, Osterholz sent me an email: "Ben, I agreed to try your Construction Management — CM — approach on COOP (Continuity of Operations), a priority enterprise-level capability, subject to available

```
Wed. 6/12/2002  11:23 PM

From: Osterholz, John, CIV, OSD-C3I [John.Osterholz@osd.mil]
Sent: Wednesday, June 12, 2002 11:23 PM
To: 'w.b.jones@verizon.net'; Osterholz, John, CIV, OSD-C3I; Elliott,
Oma, Ms, OSD-C3I
Cc: Harney, Ruby, CIV, OSD-C3I; Cross, Paula, CIV, OSD-C3I; Cherry,
Marian, , OSD-C3I
Subject: Re: Face Time

Ben,

I agreed to try your CM approach on COOP, a priority enterprise level
capability, subject to available funding. You are the only person that I
found who has correctly pointed out that CM is a engineered process and not
an engineering tool suite.

We are continually awaiting our appropriations collegues to release the O2
Counter Terrorism supplemental funding request. I don't believe in doing
anything else until then as it could only serve to divert you from persuing
your overall business plan and other near term leads. Its not fair to do
otherwise at this time.

Regards,
Jlo
```

funding. You are the only person that I found who has correctly pointed out that CM is an engineered process and not an engineering tool suite." Read that again. The only person.

Every other contractor, every other vendor, every other solution provider had been selling tools — software packages, network management systems, monitoring platforms. They were selling hammers and asking the Pentagon to figure out how to build the house.

I was selling the blueprint. The engineered process. The systematic approach that would actually solve the problem, regardless of which tools you used.

COOP — Continuity of Operations. The systems that keep the government running during emergencies. After 9/11, nothing was more critical. And Rumsfeld's chief engineer had agreed to pilot **Network Law & Order** on those priority systems. Then came the last paragraph:

We are continually awaiting our appropriations colleagues to release the O2 Counter Terrorism supplemental funding request. I don't believe in doing anything else until then, as it could only serve to divert you from pursuing your overall business plan and other near-term leads. It's not fair to do otherwise at this time.

June 2002. Nine months after 9/11. Counter-terrorism funding was flowing in billions. The Pentagon could barely spend money fast enough.

But not for NCM. Not for the "priority enterprise level capability" that Rumsfeld's chief engineer had agreed to pilot.

"Subject to available funding" is Washington-speak for "The decision has been made somewhere else, by someone else, for reasons we can't discuss." The funding never came.

Lieutenant General Al Edmonds, former Director of DISA – The NMCI Celebration

An office desk listed on its side in the fountain pool. Confetti everywhere. Was that a lady's high-heeled shoe floating in the water?

Electronic Data Systems (EDS) had just won the Navy-Marine Corps Intranet contract, officially awarded on October 6, 2000. Billions of dollars. Multi-year. The kind of deal that makes careers and breaks competitors.

At the helm of EDS Government Services: Lieutenant General Albert Edmonds, Air Force, retired. Before EDS, he'd been Director of DISA — the three-star command that runs telecommunications for the entire Department of Defense.

"Ben, stick close to Intae Kim," General Edmonds advised as Cliff and I walked into the aftermath of what must have been one hell of a party.

Intae Kim was his Chief Engineer — the technical gatekeeper. "Stick close to Intae Kim" meant convince him, and we'd have EDS's backing on the largest military network contract in history.

Kim was thorough. He took two days under the hood and kicked tires. Methodical. He asked the kind of questions only someone who'd actually managed large-scale networks would ask. How does NLO handle version control? How do you prevent configuration drift? How do you baseline complex topologies? What does the audit trail look like?

We answered every question. Showed him the proof of concept. Walked him through real-world scenarios. By the end of the second day, Kim was nodding. “This works. This actually works.”

General Edmonds didn’t just endorse NLO after that session. He became an advocate. He actively marketed the solution to anyone who’d listen.

Here was a three-star general, Air Force, former Director of DISA, telling the defense industry: “This is what we need. This solves the configuration management problem we’ve had for decades.” The truth: Network Configuration Management was so much of an afterthought on the contract that shoehorning in NCM would drastically affect the “seat price” of the contract. I recall Intae Kim and Edmonds making several attempts to highlight the need. In any rational system, that would have been enough. It wasn’t.

Lieutenant General Harry Raduege, former Director of DISA – the Panel of Experts

“Ben, good job, but you gotta respect the acquisition process,” said Raduege as I was packing up after the Panel of Experts briefing. Lieutenant General Harry Raduege was Director of DISA when we briefed there — the same three-star command that runs telecommunications for the entire Department of Defense. DISA is the AT&T of the military. When DISA’s director evaluates telecommunications technology, that’s the authoritative voice.

General Raduege didn’t just look at NLO casually. He was required to host a full panel of experts briefing at DISA headquarters.

A panel of experts isn’t a courtesy meeting. It’s a formal technical evaluation. You bring in subject matter experts from every theatre of operations — engineers, program managers, operational commanders — and you systematically assess whether a technology does what it claims to do. It's rigorous. It's thorough. It's how the military is supposed to evaluate solutions before committing resources.

We briefed the panel. Answered their questions. Demonstrated the approach. The panel understood what we were offering. DISA — the organization that would benefit most from systematic NCM — had seen NLO evaluated by their own experts.

General Raduege never publicly endorsed it. I distinctly recall looking around the large conference room full of expressionless faces and saying, "I'm confused. Here I am, a small business. I thought y'all were supposed to help me?"

Brigadier General Velma Richardson, Commander – US Army Network Enterprise Technology Command (NETCOM) – Fort Huachuca, Arizona

Brigadier General Velma Richardson made history as the first Black female commander of NETCOM — the Army's newly formed Network Enterprise Technology Command. NETCOM's mission: operate and defend Army networks worldwide.

You cannot defend networks you cannot identify. Configuration management isn't optional for NETCOM — it's foundational to their entire mission.

General Richardson understood this. But she didn't just look at NLO and decide she liked it. She did it the right way: She advertised a competitive solicitation online. A competitive solicitation means:

Published requirements.

Open competition.

Multiple vendors can bid.

Formal evaluation criteria.

Selection based on merit.

General Richardson selected **Network** Law & **Order** for NETCOM consideration. We traveled to Fort Huachuca, Arizona, to brief the NETCOM staff on implementation. The briefing room was filled beyond capacity. We had to move to the cafeteria to accommodate the overflow crowd. The interest was real. The need was obvious. The solution had been selected through fair competition. Her aide, then Major Keith Nicolette, did everything he could to get it funded. We were never funded.

A one-star general — the first commander of the Army's network defense organization — ran a competitive procurement process, selected a solution based on merit, and the funding never materialized. Keith Nicolette tried. General Richardson had made her selection. But somewhere between military requirement and actual appropriation, the money disappeared.

A competitive solicitation — the most transparent, merit-based way to procure military technology — was overruled. Not with technical justification. Not with a competing solution. Just... no funding.

The Pattern

Throughout these briefings, one thing kept happening. Engineers would ask technical questions — detailed, complex questions about how NLO worked — and they'd direct those questions to Cliff's son.

Cliff's son, who is White, was our support staffer. He operated the slides during presentations. Not an engineer. Not the inventor. Just the guy running the PowerPoint. But again and again, military engineers would look past me and ask him the technical questions.

Finally, at one briefing, he'd had enough. "Excuse me, sir, but why do you ask me questions? I am just the slide operator. He's the inventor." We got that all the time.

When Generals Don't Matter

Let me summarize what happened between 1999 and 2012:

One Four-Star Chairman of the Joint Chiefs of Staff ordered his J6 to brief **Network** Law & **Order**.

One Four-Star Logistics Commander assembled every CIO

under his command to evaluate it.

Five Three-Star Generals encountered it:

Keith Kellogg (J6, briefed on Chairman's order)

Harry Raduege (DISA Director, hosted panel of experts)

Alonzo T. Short (former DISA Director, validated at Osterholz meeting)

Albert Edmonds (former DISA Director, EDS president, became public advocate)

Bernie Schoch (Deputy Commander DISA, appeared to be naïve about the acquisition process)

One One-Star General (Velma Richardson, NETCOM) ran a competitive solicitation and selected it.

The Pentagon's Chief Engineer recognized me as "The only person who has correctly pointed out that CM is an engineered process" and agreed to pilot it (NLO) on priority systems.

Two of Those Generals Publicly Advocated for NLO: Edmonds and Short. Both retired. Both former DISA Directors. Both Black.

The active-duty generals — the ones who could actually fund and deploy it — stayed silent. Not because they didn't believe in it. The Chairman ordered attention to it. Lyles assembled his entire leadership to see it. Kellogg took the Lyles briefing the week after Shelton ordered it. Raduege hosted a formal panel of experts. Richardson selected it in an open competition.

But none of the active-duty generals could publicly champion it without career consequences. Because the tribe — the National Security Telecommunications Advisory Committee (NSTAC), the Telecommunications Industry Association (TIA), the entrenched contractors spending over $100 million annually lobbying Congress — had made it career suicide to advocate for a solution from a Black-owned small business that would bring transparency and accountability to a trillion-dollar industry built on opacity.

The tribe couldn't attack the technology. The Pentagon's chief engineer had validated it in writing. The tribe couldn't meet the military's needs. Post-9/11, after the Pentagon couldn't rebuild its own infrastructure because baselines didn't exist, the requirement was undeniable.

The tribe couldn't attack the generals' credentials. These were the Chairman of the Joint Chiefs, current and former DISA Directors, the Army's network defense commander, and the J6 for all military communications.

So the tribe attacked the funding. And they weaponized career consequences to ensure no active-duty general could fight back without risking their stars.

Richardson's competitive selection got defunded. Osterholz's pilot "awaited funding" that never materialized. Kellogg stayed silent. Raduege stayed silent. Lyles stayed silent.

Only the retired generals — the ones whose careers were already over — felt safe enough to publicly endorse a Black inventor's solution to a problem the entire military recognized was critical.

The Racism

This is the racism of cybersecurity. Not burning crosses. Not explicit slurs. Not "we don't fund Black businesses."

Systemic weaponization of career consequences to prevent anyone — Black or White, general or civilian, Democrat or Republican — from championing a solution that would disrupt a trillion-dollar industry's business model.

The Chairman orders it? Ignore the order. A competitive selection chooses it? Defund the selection. The Pentagon's chief engineer validates it? "Await funding" forever. Former DISA Directors endorse it publicly? They're retired, they don't matter. Active-duty generals see it and understand it? Career suicide to champion it.

Even when you brief YOUR technology, engineers ask the White slide operator the technical questions. The technology dies, not from direct rejection, but from systematic suffocation of every path to deployment. And the tribe does this without ever having to say the quiet part out loud.

Because if they said it explicitly — "We're blocking this because the inventor is Black"— there would be consequences. Investigations. GAO reports. Congressional hearings.

But if they just make the funding disappear. If they just ensure that every general who might champion it faces career consequences. If they just keep asking the White guy about the Black inventor's technology. If they just keep saying "Very interesting, we'll take it under advisement," while the appropriations process mysteriously fails to materialize. Then they'd get to keep their jobs.

Chapter 3 reflects the author's firsthand experiences and professional interactions with government and military officials. References to meetings, briefings, or expressions of interest describe professional engagement and evaluation of concepts, not endorsements, approvals, or recommendations of any commercial product, service, or company. Any interpretations are the author's personal perspective.

CHAPTER 4

THE FEW; THE PROUD

"Les, who are those White boys?" There were three of them. They had that nice, all wood furniture, nice carpets, but they were seldom seen. They traveled the world, and few knew what they did. They were pedigree; I was a mutt.

The CCITT — the Consultative Committee for International Telegraphy and Telephony — was where telecommunications standards were made. These were the engineers who decided how networks would communicate across borders, which protocols would become mandatory, and which technologies would be blessed with international adoption.

Overwhelmingly White and male at the time. All from the major carriers and equipment manufacturers. A Black engineer in that CCITT ecosystem was like a rare bird sighting — technically possible, but something no one expected to actually see. It simply didn't happen. The standards that would govern global telecommunications for decades were being written by a tight professional circle that looked exactly like you'd expect: exclusive, insular, and utterly convinced of its own expertise.

"Ben, don't worry about it," said Les.

These weren't just engineers. They were high priests of a technological religion, traveling from Geneva to Tokyo to Washington, making decisions that would shape trillion-dollar industries. And if you weren't part of their club — if you hadn't come up through AT&T or Bell Labs or one of the approved European carriers — you didn't get a seat at that table.

I wasn't part of their club. But I did see our GTE representative get out of the taxi from the parking lot, all tanned in the middle of January, looking for a coat.

But I thought maybe my technology would speak for itself. I thought maybe if I could get them to understand that standards needed a human intelligence component. I understood their blowback. I, being a minority, and suggesting labor-intensive activities, sounded like a social plan for job creation. Having technology doesn't make you smart. These prima donna engineers understood one thing I never contemplated. They understood standards did not translate to capital. I was wrong about how power actually works.

How Power Actually Works – Highway Helpers

The Pentagon Renovation Program meeting should have been my first clue. Major defense initiative. Rebuilding Pentagon infrastructure after 9/11. Billions of dollars at stake. Government officials chairing the meeting, presumably setting requirements and directing contractors.

I watched one guy put his foot down. Firm. Authoritative. Telling everyone in the room what the rules would be, how the process would work, and what the requirements were.

I leaned over and complimented him afterward. "Good for you, taking charge like that. Someone needed to tell us contractors how this was going to work." He smiled. "I'm the contractor."

Not a government official. A contractor. He was setting the tone, and everyone in the room — including the actual government officials — was following his lead.

That's when I started to understand. This wasn't government procurement with contractors bidding on requirements. This was contractors setting requirements and government officials formally approving them.

This system was not hidden in secrecy. It was operating in plain sight. You just had to know what you were looking at.

The Private Opinion

Beth Clay was my White lobbyist. Smart, connected, and knew how to navigate Capitol Hill and the Pentagon. She'd arranged meetings, made introductions, tried to open doors that I couldn't open myself.

After one of her meetings with John Osterholz — Secretary of Defense's chief engineer, the man who'd sent me the email calling me "the only person" who correctly understood configuration management — she pulled me aside. "Ben, I need to tell you something. John says everyone thinks you're a charlatan." I stood there, trying to process that.

Osterholz had written, in an official email at 11:23 PM, that I was "the only person that I found who has correctly pointed out that CM is an engineered process and not an engineering tool suite." He'd agreed to pilot NLO on priority systems. He'd validated the approach in writing. But privately, he said people were saying I was a charlatan Public validation. Private assassination.

That's how the machine works. You can't attack the technology directly — too many generals had endorsed it, too many briefings had validated it. But you can whisper. You can plant doubt. You can make sure that anyone considering funding knows: "Everyone thinks he's a charlatan." Everyone? Or just the tribe that needed me to be one?

CDO Technologies — The Workaround

I was facing foreclosure on my house, 20 years of work, a Howard engineering degree, 15 years at GTE Government Systems, a pre-grant publication on **Network** Law & **Order**, briefings to generals, and validation from the Pentagon's chief engineer. And I was about to lose my home.

So I got creative. If they wouldn't buy from a Black inventor, maybe they'd buy from a small Black-owned and well-established business.

I researched and found CDO Technologies, a minority-owned company headquartered in Dayton, Ohio — coincidentally near Wright-Patterson Air Force Base, where we'd briefed General Lyles a year earlier. By 1999, CDO Technologies had become one of the leading small technology contractors in the Dayton defense technology corridor. Established. Credible.

We entered into a formal agreement. I would lease them **Network** Law & **Order** for $28,000 — just enough to stave off foreclosure. I'd train CDO on NLO, give them my DISA contacts, and see if they could sell what I couldn't.

CDO didn't know my financial situation. They saw a business opportunity: license proven technology, sell it to DoD, take their cut. Standard business deal. DISA leaped at the chance.

Suddenly, the same organization that had been dragging its feet for years was eager to move forward. They took CDO's brief. Internal emails started circulating requesting funds to proceed. The enthusiasm was real. Finally, I thought. Finally, it's going to happen.

Then CDO had to reveal product ownership. A standard requirement in government contracting — you have to disclose intellectual property rights, licensing agreements, and who actually owns what you're selling.

Once they found out it was me, the funding disappeared. Same technology. Same need. Same DISA personnel who'd been enthusiastic about the CDO brief. But now that they knew a Black man held the license? Silence.

Patent Pending! Oh Shit. The Truth Spoken Out Loud

I kept wondering why, at every briefing, we'd get sidetracked about the status of our patent submission. On the first snow day of the new year, I had a visit from an old colleague — a White government engineer I'd worked with in Hawaii. We'd stayed in touch over the years. I'd done favors for him back when I was a GTE contractor — procuring T1 circuits, solving technical problems, the kind of professional courtesy that builds relationships.

He came to visit me after the CDO deal fell apart. "Ben, the problem isn't your technology, it's your license." He said it plainly, like he was explaining something obvious that I should have understood already. "No way in hell will they buy your license for so large a project. He expressed the view that such power was unlikely to be placed in Black ownership."

There it was. Spoken out loud by someone who knew how the machine worked, who'd seen it operate from the inside, who understood exactly why six generals' endorsements and a Chairman's order weren't enough. "… Not giving no Black man that kind of power."

The power wasn't in the technology. The power was in the licensing revenue, the control over a critical capability, the ability to set terms for a trillion-dollar industry, and they weren't giving that to a Black man.

He wasn't being cruel. He was being honest. He'd seen how the system worked. He was trying to help me understand what I was up against.

The machine didn't need to say it in official memos. It didn't need to document the racism. It just needed to ensure that when ownership was revealed, the funding disappeared. When licenses were Black-owned, the enthusiasm vanished. When power might shift to someone outside the club, the machine said no.

Eddie Huff Got Me in this Mess: The Revolving Door

Eddie Huff was Senator Tom Coburn of Oklahoma's friend. I met him at a roundtable on Capitol Hill, and we became friends. Eddie introduced me to Senator Coburn, who, in turn, introduced me to the Senate's proclaimed IT engineer, on staff for Senator Tom Coburn. Good guy. Smart. And like many of us, he understood some of the evolving technology, not all of it. We'd worked together on telecommunications oversight — the senator was grilling carriers about network outages, and the IT engineer needed to understand how networks were actually configured and managed to ask the right questions. The senator wanted me to draft the questions that would get to the bottom of the connection issues. I was asked to have the senator ask me directly when Mike Swartz, Coburn's Chief of Staff, intervened. "Ben, we need some expertise in the shadows, and he just ain't up to it." Over the next week, I designed questions and reviewed them with the IT engineer so he could explain them to the senator. What a thrill to hear your questions being asked by a senator and the paper-shuffling and head-scratching responses from the witnesses, all from the major carriers.

Then the IT engineer got hired by the Department of Homeland

Security (DHS). Big job. Significant promotion. Good for him.

Senator Coburn suspected something. Mike Swartz, his Chief of Staff, called me. Saturday morning. Maryland mall. Away from the office, away from official channels.

"Did the IT engineer getting that high-powered job have anything to do with the withering questions?" Mike asked. I told him the truth. "That's how the system works, you know that better than me."

Mike nodded. He knew. The senator knew. Train the government's technical expert, give him inside knowledge of your operations, then hire him away before the investigation gets serious. The questions weaken because the person writing them now has a job offer on the table — or has already taken the job and doesn't want to burn bridges with his new employer. It's not illegal. It's not even unusual. It's just how the system tends to protect its own interests.

I cannot say that this was the case with the Senate's IT engineer. He was, in fact, trying to do the right thing in defense of this country. Going to DHS should have enabled him to do more. Maybe it did.

Mike and the senator knew they owed me an ask. I'd coached John. I'd helped them understand the telecommunications nuances within the change management industry. So I asked.

The Lie

My ask was simple: have DISA send a senior representative to tell Senator Coburn's staff the truth about DISA's Network Configuration Management capability.

DISA sent a senior official who oversaw the DSN (Defense Switched Network), DRSN (Defense Red Switched Network), and DDN (Defense Data Network) — the major DoD communications systems. If anyone knew the actual state of DISA's configuration management, it was someone at that level.

That official came to Capitol Hill to brief Mike Swartz and the senator's staff. And the information provided was, in my assessment, inaccurate: DISA claimed NCM capability. Had it covered. No need for outside solutions. Everything under control.

It wasn't true. DISA didn't have a systematic NCM capability. That's why General Raduege had hosted a panel of experts to evaluate NLO. That's why DISA had "leaped at the chance" when CDO presented the same technology. That's why the Pentagon couldn't rebuild its infrastructure after 9/11 without a $400 million program — because baseline configurations didn't exist. But DISA told Congress they had it covered.

And that was that. My one ask, the favor owed by a senator and his chief of staff, was spent on a lie that ensured NLO funding wouldn't come through congressional pressure.

Mike Swartz passed away while I was going through my own cancer treatment. His wife informed me after I had recovered. I did not know Mike was ill. We had both kept our conditions private, the way men of a certain generation tend to do. Two men fighting the same battle in silence, both trying to do right by their country, neither knowing the other was in the fight of his life.

The Machine Revealed

NSTAC was created by Executive Order 12382 in September 1982. It's a panel of industry executives who advise the President on national security, emergency preparedness, and communications policy.

Sounds reasonable, right? Government getting expert advice from industry leaders?

Except that the meetings are almost always closed to the public. National security, they say. And NSTAC's website claims that documents and records aren't subject to the Freedom of Information Act because it's not technically a federal "agency."

That structure deserves more than a raised eyebrow. It reflects something fundamental about how America handles telecommunications: roughly 70 percent of the nation's infrastructure is privately owned, built and maintained by companies that answer to shareholders. The remaining 30 percent is government-held. But the government carries 100 percent of the responsibility for protecting the country when those networks go down or get compromised. Telecommunications is inherently governmental in its consequences, even when it's privately operated. NSTAC is one of the mechanisms America invented to bridge that gap. It is imperfect, like most bridges. But it exists for a reason.

TIA spent over $100 million lobbying Capitol Hill in 2006 alone.

Cliff and I spent $64.08 that year. Most of it on parking.

TIA represents telecommunications networks and control systems — the standards, the interoperability frameworks, the technical backbone that makes it possible for networks to function across carriers, platforms, and national boundaries. NCTA and the cable industry built the broadband infrastructure that now serves as primary telecommunications for tens of millions of Americans. NSTAC brings the executives of these industries into a room with national security implications on the table. Together, they sit at the intersection of private capital and public safety in a way no other country has quite replicated. They operate within a system they did not design — one where profit and national security share the same wires — and they do what such organizations do: They protect what they know, they shape what they can, and they move at the speed of capital rather than the speed of threat. That is not a moral failure. It is the predictable behavior of institutions operating under structural constraints that nobody has honestly resolved. In my experience, that structure played out at every level:

International: CCITT sets standards with an all-White club of approved engineers.

Policy: NSTAC advises the President in secret meetings.

Lobbying: TIA, NCTA, and the cable industry spend hundreds of millions annually ensuring Congress understands their infrastructure priorities.

Procurement: Contractors shape requirements, and government officials formally approve them.

Personnel: There is a revolving door between government oversight and industry jobs.

Execution: When ownership is revealed, funding disappears.

You can't fight the machine at one level because it operates at all levels simultaneously. You brief a general? The machine attacks the funding. You get Panel of Experts validation: The machine whispers "charlatan." You work around it with a Black-owned certified business: The machine requires ownership disclosure, and funding vanishes. You train congressional staff: The machine hires them away. You ask for the truth: The machine sends someone to Congress to lie.

And when all else fails, the machine has people — honest ones, even — who will tell you the truth: "He expressed the view that such power was unlikely to be placed in Black ownership."

Not because of technology. Not because of qualifications. Not because six generals validated the claimed benefits, a Chairman's order, or Panel of Experts validation.

Because power is what the machine protects. And power doesn't flow to people outside the club.

The machine is perfectly designed to do what it does: extract maximum revenue from government contracts while preventing transparency, accountability, or competition that might threaten the trillion-dollar business model built on complexity, opacity, and the economics of a system that may be charging $1.76 for every $0.089 of Information Services delivered.

It's not a conspiracy. Conspiracies require secrecy. The machine operates in plain sight. You just have to know what you're looking at. And once you see it, you can't unsee it.

Chapter 4 reflects the author's firsthand experiences, interpretations, and opinions formed during his professional career. Descriptions of institutions, meetings, and individuals are presented as personal perspective and historical context. Statements regarding influence, access, or decision-making describe perceived institutional dynamics and are not assertions of unlawful conduct, discriminatory intent, or coordinated action by any specific individual or organization.

CHAPTER 5

WHO RUNS BARTERTOWN?

"They will never admit it, but they get goose bumps seeing in ultra-high definition the license plate of the enemy they are tracking. Wow, right out of Hollywood. Master Blaster provides these benefits to a select group, allowing them to charge the majority $1.76 for every $0.089 of Information Services delivered." Ben Jones

Let that phone go dead or the computer go "blue screen." Wait on hold for an hour and when someone finally comes online tell them you are in control and the value of your time. See how long you wait on hold.

In the movie *Mad Max Beyond Thunderdome*, there's a post-apocalyptic trading settlement called Bartertown. Everyone thinks Aunty Entity runs it — she sits on the throne, makes pronouncements, enforces the rules. But the real power lies underground with Master Blaster, who controls the energy that keeps Bartertown running. Aunty Entity can posture all she wants, but without Master Blaster's cooperation, nothing works.

The question that echoes through the movie — "Who runs Bartertown?" — isn't asking who appears to be in charge. It's asking who actually controls the power.

That's the question I should have been asking in 1999 when I started briefing generals.

Who runs defense procurement? The generals who issue requirements? The Pentagon officials who evaluate solutions? The congressional committees that appropriate funds? The President who sets national security priorities?

Or the contractors who write the requirements, the industry executives who advise in secret, the lobbyists who spend $100 million annually, or the machine that makes careers and funding disappear when anyone challenges the system?

I thought the government ran Bartertown.

I was wrong. The government is addicted to Bartertown. The government likes to turn a switch and watch bombs drop on a target 5,000 miles away in stunning 4K high definition. Master Blaster knows this but requires frequent reminding.

The Truth Spoken Out Loud

That encounter with my colleague from Hawaii — his plain statement about the license, about where power would and would not be placed — was not the voice of Master Blaster. It was someone who understood Bartertown well enough to tell me the truth about it.

The power was never in the technology. It was in who held the license — and what that meant for who controlled a critical capability in a trillion-dollar industry.

And they weren't giving that to a Black man.

Chapter Four showed you how the machine operates: the NSTAC structure, the CCITT club, the $64.08 against TIA's $100 million, the revolving door, and the plain-spoken truth from the man who came to visit after the CDO deal fell apart. You now know the mechanism. What Bartertown adds is the metaphor that makes it visible —

because some truths only land when you see them from the right angle.

Aunty Entity sits on the throne and makes speeches. Master Blaster controls whether the lights turn on.

And once you understand how Bartertown really works, the next question is not who to blame. It is who can fix it.

The consequences were real. The 650,000 jobs that NLO would have created — jobs that could not be outsourced, jobs for young people with the skills to do them — never materialized.

Not because the technology didn't work.

Not because the military didn't need it.

Not because the economics didn't make sense.

Because Master Blaster said no.

And in Bartertown, when Master Blaster cuts the power, even Aunty Entity sits in the dark.

Even generals can't make the lights turn on.

LEGAL NOTICES AND DISCLAIMERS

CHAPTER 6

LIVES LOST

"Lone Survivor"

Master Blaster runs Bartertown. NSTAC and TIA influence the power. Contractors set requirements. Generals get overruled. Black inventors get blocked.

The author views this as a serious failure of process — systematic exclusion, procurement bias, and a large system protecting its own interests. But there's another crime. The one measured in lives instead of dollars.

Operation Red Wings

June 28, 2005. Four Navy SEALs were inserted into the mountains of Afghanistan's Kunar Province for a reconnaissance mission. Their target: a high-value Taliban leader.

The team: Lieutenant Michael Murphy, Gunner's Mate Second Class Danny Dietz, Sonar Technician Second Class Matthew Axelson, and Hospital Corpsman Second Class Marcus Luttrell.

The movie *Lone Survivor* tells what happened next. The team encountered goat herders. Faced an impossible choice: kill civilians or let them go, knowing they'd alert the Taliban. They let them go. Within an hour, they were surrounded by thirty to eighty Taliban fighters.

The firefight was brutal. Outnumbered and outgunned, the SEALs fought from high ground, trying desperately to call for help. They couldn't get through.

Radio communications failed. Satellite communications failed. Cell phones failed. The mountainous terrain, the remote location, the limitations of their equipment in that specific environment — all conspired to leave four of America's most elite warriors unable to call for the help that was waiting to respond.

Lieutenant Murphy exposed himself to enemy fire, climbed to higher ground seeking better reception, and finally got a brief transmission through to headquarters. He was shot and killed moments later.

A quick reaction force was launched to rescue the team. A Chinook helicopter carrying eight SEALs and eight Army Night Stalkers was hit by an RPG. All 16 died.

By the end, 19 American warriors were dead. Marcus Luttrell was the lone survivor, rescued by Afghan villagers who protected him until US forces could extract him.

The heroism of that day is undeniable. Lieutenant Murphy received the Medal of Honor. The courage, the sacrifice, the brotherhood — all of it real, all of it honored.

But there is a question the author believes was insufficiently examined.

The Question

Why were four Navy SEALs inserted into an area where they couldn't communicate? This wasn't 1945. This was 2005. We had satellites. We had encrypted radios. We had technology that should have ensured those warriors could call for help if they needed it.

In the author's assessment, the issue was not solely the equipment. The author believes a digital site survey may not have been conducted before the mission. Basement level tactical NCM. The most basic application of the science I'd been trying to get DoD to adopt for six years.

A digital site survey could potentially have revealed that the area had poor cell reception, limited satellite visibility due to mountainous terrain, and radio propagation challenges. That information would have informed the mission planning. Different insertion point. Different communications protocols. Different backup plans.

Or maybe a different assessment of whether four men could operate in that environment without reliable communications.

Digital site surveys are not always conducted as standard practice. We don't systematically assess network capability before deploying forces. We don't treat communications infrastructure with the same rigor we treat weapons, ammunition, and supplies.

Because we don't have systematic NCM. Not at the strategic level. Not at the tactical level. Not even at the basement level that would tell you: "Don't put SEALs in a place where they can't call for help."

What NCM May Have Prevented

This isn't speculation. This is what tactical NCM does: It maps network capability — cell towers, satellite coverage, radio propagation, terrain interference — and provides commanders with accurate information about where their forces can and cannot communicate.

It's not exotic technology. It's basic engineering. Survey the environment. Map the infrastructure. Identify the gaps. Plan accordingly.

We do this for civilian cell phone networks. Drive around, measure signal strength, identify dead zones, and place towers accordingly. It's so routine that telecom companies have automated systems for it.

But we don't do it systematically for military operations. Because doing it right requires NCM — the same discipline that DISA didn't want to adopt, that NSTAC and TIA blocked, that Master Blaster said no to because it would bring transparency and accountability to a trillion-dollar industry.

From the author's perspective, the team may have entered the mountains without full awareness of communications limitations. And when they desperately needed help, they couldn't call for it. And Lieutenant Murphy had to climb higher, expose himself to enemy fire, try to get line-of-sight to a satellite, just to send the message that should have been a routine radio call: 19 warriors died that day.

Some of those deaths were inevitable. War is brutal. Courage doesn't guarantee survival. Even perfect communications couldn't have saved everyone once the firefight started.

But some of those deaths might have been preventable. If we'd known the communications environment before insertion. If we'd planned for the dead zone. If we'd had protocols for operating without reliable communications. If we'd asked the basic question: "Can they talk to us from there?"

That's what NCM does. It asks the boring, unglamorous, bureaucratic question that saves lives: "What's the network configuration?"

The Pattern

Operation Red Wings wasn't an isolated incident. It's a pattern. 9/11: The Pentagon can't rebuild its infrastructure because nobody maintained configuration baselines. A $400 million program is required because we didn't spend $15 million on NCM beforehand.

Hurricane Katrina: Emergency responders from different agencies couldn't communicate because nobody had mapped interoperability requirements across the patchwork of networks.

Every major disaster: Communications fail because we don't systematically manage network configurations, don't maintain baselines, and don't know what's actually deployed versus what's supposed to be deployed.

Every day, cybersecurity vulnerabilities hide in the gaps between documentation and reality because we can't secure what we can't identify.

The price of not having NCM isn't just wasted money. It's not just political careers. It's not just a Black inventor unable to get his solution funded despite six generals validating its claimed benefits. The human cost can be severe.

The Crime Completed

Master Blaster runs Bartertown. The machine protects the trillion-dollar business model built on opacity and complexity. NSTAC and TIA ensure that solutions bringing transparency and accountability never get funded. That assurance is not programmatically diabolical or intentional, rather the consequence of focused connectivity solutions devoid of a business process.

The consequence is that warriors die in the mountains, unable to call for help, because nobody bothered to do a digital site survey. The author views that as the central failure.

Not just what they did to me. Not just the racism of systematically excluding Black innovation. Not just the arrogance of contractors setting government requirements.

The crime is what they stole from the country: the capability that would have given four Navy SEALs a fighting chance to call for help. The infrastructure accountability that would have let the Pentagon rebuild after 9/11. The network security that would have protected critical systems from adversaries hiding in un-auditable configurations.

Six generals understood this. The Chairman of the Joint Chiefs ordered attention to it. The Pentagon's chief engineer validated it in writing. A competitive solicitation selected it.

And Master Blaster said no, because "They're not giving no Black man that kind of power." So the lights stayed off. The networks stayed unmanaged. The configurations stayed undocumented. And warriors died in communication dead zones that basement-level NCM would have identified before insertion.

That's not bureaucratic failure. That's not just racism. That's not just corruption. The author views this as a grave systemic shortfall with life-and-death consequences.

But the warriors who died in operations hampered by communications failures? They'd been paying the price for years. We just didn't connect the dots. Until now.

Chapter 6 reflects the author's professional opinions and personal interpretations regarding historical events and publicly documented operations. References to military incidents, planning decisions, or communications limitations are offered as analytical perspective, not as definitive conclusions or allegations of fault, negligence, or misconduct by any individual, unit, or agency.

"As with every cyber hearing this committee has held in recent years, we've heard how the lack of strategy and policy continues to undermine the development of a meaningful deterrence in cyberspace, the threat is growing. Yet we reamin stuck in a defensive crouch forced to handle every event on a case-by-case basis and woefully unprepared to address these threats."

—Senator John S. McCain,
Chairman Senate Armed Services Committee Hearing:
Cyber Policy, Strategy, and Organization,
May 11, 2017 (S. Hrg. 115-181)

THE SOLUTION

Chapters 7-11

CHAPTER 7

BACKDOOR BISON

The Howard University mascot is a Bison, and anyone who graduates from Howard University is a Bison, i.e., "Let's Go Bisons!"

I am a Backdoor Bison as well as being the 1983 Class Goat in the Louis K. Downing School of Electrical Engineering at Howard University. The Class Goat is the graduating student with the lowest Grade Point Average (GPA). The front door or normal way to enter a university is by providing an application to that university which includes college transcripts, GPA, SAT scores, a personal letter of interest, and letters of recommendation. I had none of the above because I did not take any college preparatory classes, as none were available for my section.

Section? Yes, in high school, we were divided into sections, usually four sections per grade level. For example, if you were in the tenth grade, you were either in 10-1, 10-2, 10-3, or 10-4. The higher the number, the "smarter" you were perceived to be. I was 10-3, meaning no hope of college but a good chance of obtaining a suitable skill. Those students in 10-1 and 10-2 classrooms were considered "college material" and were exposed to college preparatory courses such as algebra and biology. So, how does a 10-3 student in a segregated high school in Fayetteville, North Carolina, become a graduate in Electrical Engineering from the Louis K. Downing School of Engineering at Howard University? Employee. That's how. In 1973, the Department of Health and Human Services (DHHS) awarded Howard University Medical System a capitation grant to provide information technology to the schools of medicine and dentistry.

A 3.5-inch closed-circuit television system, complete with broadcast studios and technical support facilities, required the recruitment of broadcast television and radio repair specialists. I landed one of two jobs, one a GS-8 and the other a GS-10. I had the credentials for the GS-10 position; the White guy had the knowledge but no credentials. He received the GS-10 slot, which was fine with me because I did not know the technical side of broadcasting. Even then, I got the feeling that Howard wanted me to have the GS-10 position because they were very impressed that I had acquired a Third Class FCC license at 14 and a First-Class FCC license at 16. I did not have any appreciation of the accomplishments nor any context by which to gauge the value of a 21-year-old with such lofty credentials. The logic was that I should have the higher pay because of the higher credential and it was at a historically Black college. The problem was I wasn't qualified to that degree of technology. A witness to the dental school hiring event said later on, "I think they were reluctant to give the top position to the new guy; it didn't look good." It became the mission of the Dean and her academic staff to ensure that I attend Howard University School of Engineering.

I can only assume their mission began with the dean insisting that I be her seatmate on the return flight from Meharry School of

Dentistry in Nashville, Tennessee. I had accompanied the dentists and teachers as support for their technical presentations to their fellow dental school colleagues. The dean, having heard about my abilities, decided to inquire as to how I was able to obtain a third and first class license at such a young age. During the flight, the dean essentially said that I had one year to enroll in engineering school. She labored to explain why a degree would be better than my current license. "One year, Ben, you have one year to demonstrate that you are working toward a degree. I am here to help you in any way I can." I told the dean that I would apply to a school at Howard during the year. She looked at me for a brief moment then went back to her reading.

A year later, I hadn't thought about the conversation with the dean when suddenly, my bi-monthly paycheck did not appear in the staff mailbox. I did find a note inviting me to come to the dean's office as soon as possible. I thought to myself, with no paycheck, the right and soonest time possible was right at the moment. Surprisingly, the dean's secretary said she anticipated my coming to see the dean. Moments later, I was in her office as she was ending a call. "Looking for something?" She said as she took an envelope out of her drawer. Good, I thought to myself, at least my money is here. Why did she have my money? She said she didn't have the authority to withhold my paycheck, but reminded me of our past conversation and my pledge to apply to school. This was when I explained that I lacked the credentials most colleges required. She handed me a note and instructed me to walk directly up the street. About halfway up, on the left side, I would be intercepted by a man identifying himself as Dr. DeLoach. "Go with him and do as he says. When you finish, see Wanda for your paycheck," she concluded.

It took me five years to complete the degree work at Howard University School of Engineering. The first year, I was behind my peers due to not being exposed to college preparatory classes (i.e., the 10-1 class).

My real engineering journey began with the help of Dr. Tepper Gill. My first calculus assignment exposed my ignorance. When asked about my homework, I quietly informed the professor that my particular math book has numerous printing errors. "Give me that book," said an exasperated professor. "Show me the printing errors."

I opened the book confident of my keen assessment and pointed out the pages having errors that placed tiny numbers above and next to other numbers. The professor said he understood and continued with the class. When I got up to leave, he handed me a sealed note instructing me to go across campus to an office in the basement of the student union labeled C.A.R., an acronym for Howard University's Career Academic Re-Enforcement program, and to give the note to the professor on staff. The C.A.R. program added to my math classes by putting me on a bus at 3:30 p.m., three times a week, to Northern Virginia Community College in Annandale, Virginia. There, I received a two-semester course on polynomials and algebra while keeping up with the other first-year engineering coursework.

My GPA plummeted, but due to my C.A.R. participation, they exercised academic discretion. Not for long, however, Dr. Gill taught me how to study mathematics. "What does the theorem say? Go to the board and write the theorem down in as nice a way as you possibly can." I thought this was the stupidest thing I had ever done. Over and over, the same repetitive exercises. Then he would say, "Okay, you can go."

Then the next afternoon, guess what? Back to the blackboard, writing the same theorems I wrote the day before. Finally, after what felt like years but was only two months of writing, he gave me a calculus problem. "Ben, identify the theorem at play here. Remember what you have written. Do you want to go back to the board?" With that threat, I stared at the problem, and the answer leaped out at me without me having to even decompose the steps to the solution. I blurted out the answer just based on the theorem I had learned. Dr. Gill almost cried. I was right.

I end this chapter by saying that years later, after graduating from Howard, my C.A.R. instructor, Dr. Tepper Gilll, and I became good friends. He will tell you that there is no better sausage dog than the ones we had at Vick's Drive-in on our way back to D.C. We looked at each other and made a U-turn back to order more!

Bison Full Circle

Chapter 7 reflects the author's personal recollections and life experiences.

Descriptions of academic placement, employment decisions, and interactions with educators or administrators are presented as firsthand perspective and context, not as allegations of discrimination, misconduct, or improper decision-making by any individual or institution.

CHAPTER 8

GEE. NO. GTE

A Real Job

"Jonesy, get your butt down to the Dean's office!" the voice seemed to yell over the phone. What now? I walked in, and there sat this big, fat White man with a cowboy hat and cowboy boots. His name was Don Sneck, program manager for the newly acquired Defense Switched Network contract at GTE Government Systems in Needham, Massachusetts. I looked at him. He looked at me. "So," he drawled, "this is your best and brightest?"

"Yep, one of them, Senior Project of the Year Award winner," replied Eugene DeLoach, Chairman of the Electrical Engineering Department at Howard University's Louis K. Downing School of Engineering.

"What gives?" I asked. "Hi. My name is Don Sneck of GTE. Ever heard of us?" "Yeah," I replied, "The airplane people." Back then, GTE ran ads touting their in-flight phone service and their sponsorship of the Super Bowl. "You know we also sponsor the Super Bowl?" he asked. I nodded.

What Don Sneck didn't say — what he would never say — was that I understood that the contract encouraged inclusion of at least one degree-holding Black engineer as a task leader. This was diversity, circa 1983. GTE was working to comply. They went to the closest Historical Black College and University (HBCU) with a recognized, acceptable engineering program: Howard University.

Bottom line, Sneck made me an offer I couldn't refuse.

By being "the best and the brightest," I had landed the Senior Project of the Year Award for the graduating class. It was a big deal because the sponsoring company gave you a prize. But more than that, it was a job. A real job. Starting at $28,000 a year and working on the largest telecommunications contract ever granted by the Department of Defense, what wasn't to like? I had no idea what I was walking into.

Ironies in the Digital Age: How Exclusion Sparked Innovation in America's Cybersecurity Landscape

In the early 1980s, the U.S. was a nation in the throes of transformation. The Civil Rights Act of 1964 and the newly emerging policies of Diversity, Equity, Inclusion, and Access (DEIA) were beginning to take root, stirring hope for a more just and equitable society. Yet, as with so many chapters in America's story, progress was layered with paradox and irony — especially in the rapidly evolving world of technology and telecommunications — and coinciding with my graduation from Howard University.

Recent grads (from left) Ngoc-Lan Nguyen, Dean Coclin, William Jones, Farley Osborne and Mike Mathews seem to hit it off during the recent CSD College hire orientation!

A Door Opens, But Only a Crack

The story of GTE Government Systems is emblematic of this era. In 1982, the government division of GTE landed what was then the largest telecommunications contract ever granted by the Department of Defense. As a graduating engineer at Howard University, I had no idea of the world around me. Outside events were shaping my opportunities in the form of requirements from the government. The contract came with a stipulation: The company had to hire one or more degree-holding minority engineers as task leaders, a move propelled by the spirit of DEIA.

But old prejudices die hard, as GTE Government Systems was headquartered in Needham, Massachusetts, in the greater Boston area. Its corporate culture traveled to the Virginia suburbs as well.

Instead of offering meaningful roles, I was assigned to what appeared to be a peripheral task. Documentation management. At the time, I perceived it as little more than a glorified library job. My classmates were getting glamorous (so I thought) assignments all over the world, and here I was being made a librarian.

But Here Lies the First Great Irony: Les Alberici. A White Man in the Hood.

I told Stan Fireston to "go to hell" and stormed out of GTE Government Systems located in Rossalyn, Virginia, in my first week.

"Who is it?" I asked. "Ben, it's Les Alberici from GTE, man, you gotta let me in!" Les Alberici from GTE! That White man in the big office who never smiled at anyone. What the hell? I opened the door, knowing he must have encountered Seymore. "Mr. Alberici, why are you here? I quit." Seymore was the building's designated drunk. He would piss himself on the steps of the building on South 17th street, and the neighbors would hold their breath and step over and around the snoring Seymore as they went to their flats. We all loved him. What no one could understand was why Seymore elected to sleep on the steps when his one-bedroom apartment was two steps away.

"Ben, you've got to go back and apologize to Mr. Fireston. Please hear me out. I understand what you think about Task Nine, but Ben, he just gave you 'keys to the kingdom.'" Les said, standing there, hat in hand. I was not about to offer him a seat. Now he knew I had higher standards being the Senior Project award winner in my class. I was wrong. He said I was lucky. Of all the tasks, my task was akin to a reporter's badge, except it was the government issuing the badge to me, which gave me access to all the other tasks. Hearing this, my brain went into "How to fix it mode." I instantly knew he was right.

Les was the number three man on the GTE/DSN contract as its controller. He was Italian, and years later, I found out he was rich. Les did not like how GTE leadership was treating me, so he decided to take me under his wing and schooled me about the ways of GTE. We started out taking scheduled coffee breaks across the street. Other GTE staffers started noticing, and within two months, we had ten people plus us at these breaks. Les and I stopped attending. The remaining staffers, all White, actually named the event the Old Fart's Club. I recall after having been let go, being invited to attend one of the club's gatherings — a club whose members could not tell you who or why it was started.

What GTE viewed as a dead-end role turned out to be the linchpin of the entire system. The documentation task required access to every blueprint, protocol, and network design — old and new. Documentation held the keys to the kingdom. This position — intended as an afterthought — embodied the "A" in DEIA: Access. It was a seat at the table, even if the host didn't realize it.

Access as Power, Exclusion as Catalyst

For 15 years, I worked on the Defense Switched Network. GTE government systems developed the DSN Information Management Support System (DIMSS), and to my knowledge, then helped create the Learning Acquisition Recognition System (LARS) — one of the country's first artificial intelligence applications for predicting network traffic patterns on DSN switches in Europe.

The final module was supposed to be NCM. That portion was never funded. To be fair, the technology had not caught up with the need, as documented in the first Network Configuration Management Circular. I wrote that circular. No one recognized me as the expert, but I was often treated as one in practice.

Despite the critical nature of my role, I perceived recurring bias in how decisions affected my role. Two months before my 15-year full vesting, I was a prime candidate for a Reduction In Force program.

On the surface, it was a setback. But in reality, it was the spark for something revolutionary: the invention of **Network** Law **& Order**.

This concept — rooted in understanding the designed and ordered network, the "law" of the system, not of the courts — would go on to shape the very foundation of how we think about and secure our digital infrastructure.

This is the second irony: Exclusion became the catalyst for innovation. The very act of sidelining talent led to the creation of a framework that would prove essential in the defense of America's critical telecommunications networks.

Chapter 8 reflects the author's personal recollections and professional experiences. Descriptions of hiring practices, workplace dynamics, and diversity policies are presented as firsthand perspective and historical context, not as allegations of unlawful discrimination or misconduct by any specific individual or organization.

CHAPTER 9

IF HERB CAN DO IT!

Center on

Religious
Counter Insurgency
for
Social & Economic
Development

Policy Procedures Practice

"Another buffet, another White guy, another last sandwich!"

Herb London

By now, you've noticed a pattern. High-level meetings. Powerful people, and me, standing at the buffet table, reaching for the last good sandwich.

This time it wasn't the Chairman of the Joint Chiefs of Staff. It was another White guy, mid-50s, making his move on the last Reuben sandwich at the Heritage Foundation lunchtime speaker series. At least I got the last pickle.

But this time, something different happened. This time, I made a bet. And when I won that bet, it proved something that would become the foundation for Jekyll 2.0:

If you can make a White conservative skeptic understand **Network** Law **& Order** in four one-hour lunch sessions, imagine what six billionaires could do with four full days.

The Bet

Being the only sore spot in an otherwise conservative policy crowd — Black face, electrical engineer, carrying a briefcase full of documentation reflecting prior briefings and evaluations — I drew prying eyes and EF Hutton ears. When I speak, people listen. Usually, because they're trying to figure out what I'm doing there.

The man next to me at the buffet table asked, "So, what brings you here?" I replied, "The buffet." He said, "Me too." We introduced ourselves. His name was Herb London, President of the Hudson Institute — one of the most influential conservative think-tanks in Washington. The kind of place where policy ideas become White House briefings, where think-tank papers become legislation, where serious people discuss serious solutions to serious problems.

Herb had that "you should be glad you're here" tone in his voice when he started talking — that casual condescension that assumes "I'm grateful for the invitation rather than here because I belong."

So I cut him off at the knees. "I tell you what," I said. "We're going in there to hear this guy talk for about an hour, and none of us will remember a thing he said. Mr. London, I bet you right here and now that if you listened to me for that same amount of time — "

Herb cut me off. "Listen to you for what?" I cut him right back. "As I was saying, if you gave me one hour for five days, which is a week, I bet you that the following Monday you'd be running to the Bush White House. Without these sandwiches."

Two of his friends standing nearby laughed. Herb's cheeks turned red. He looked at me like I'd just insulted his mother. "So, the deal is you're going to MAKE me go running to the Bush White House if I give you five one-hour listening sessions?" "That's right," I responded.

"So you're not convincing me. And I don't want to go to the White House. What do I get out of it?" I looked him dead in the eye. "Mr. London, you will get the gratitude of a grateful nation." The crowd around us laughed. We went into the boring session. Upon leaving, Herb slipped me a note with his business card. "It's on. Next week?" It was indeed on.

From: Herb London [mailto:Herb@hudson.org]
Sent: Thursday, September 06, 2007 8:01 PM
To: Ben Jones
Subject: RE: Lesson 1 - Think DDP

I think I've got it . Thank you

Dr. Herbert London
President
Hudson Institute
90 Broad Street
New York, NY 10004
212-232-8722 *direct*
212-232-8725 *fax*

From: Ben Jones [mailto:wb.jones@verizon.net]
Sent: Thu 9/6/2007 11:56 AM
To: Herb London
Subject: Lesson 1 - Think DDP

Herb,

David Walker is the U.S. Comptroller General and head of the General Accounting Office (GAO). I would think he is a very credible person. One must assume that if General Walker agrees with me, then perhaps I may be correct. (This is today's lesson)

)/10/2007

Five One-Hour Lunch Sessions Minus One

I gave Herb a white paper first — something to review before we started. Two weeks later, we began the five one-hour sessions. Just lunch meetings. One hour each. Five days.

This was 2007. I wasn't yet presenting **Network** Law **& Order** as "the new oil." I didn't have Chapter Two's framework yet. I was presenting it as a jobs program for middle skills and STEM — a way to train hundreds of thousands of young people in network security while solving critical infrastructure vulnerabilities. We agreed on the session time and location. I asked for a week's delay to prepare context material for our NLO sessions. That white paper, **Network** Law **& Order** is for sale at www.networklawandorder.com/NLO Store.

Each session, Herb took notes. Asked questions. Pushed back on assumptions. Demanded evidence. I provided documentation — emails, photographs, briefing materials, documentation reflecting prior briefings and evaluations.

By Session Three, he stopped interrupting. By Session Four, he stopped taking notes. He just listened. There was no session five.

I Won the Bet

That Friday, Herb's secretary handed me a note. "You won. He is at the Bush White House as we speak. He told me to tell you that you didn't need five hours."

Herb London — then President of the Hudson Institute — subsequently visited the White House, consistent with what he later conveyed to me.

Four one-hour lunch sessions. That's all it took. Not five full days. Not five intensive briefings. Five lunch meetings where he could still get back to his office for afternoon appointments.

And he understood it completely. He understood the market. He understood the workforce solution. He understood the national security implications. He understood the debt reduction potential. He understood why this needed to happen immediately.

According to Herb London, the White House requested that he

brief Homeland Security leadership. A senior official — female at the time — needed to hear this directly.

Herb had absorbed **Network** Law **& Order** in four hours total and was already discussing potential meetings at senior levels of government.

That's when I knew the concept worked. Not just the technology. Not just the business model. The communicability of it. If you could get a skeptical White conservative to understand it in four lunch hours, you could convince anyone. The ride back was filled with mixed emotions. Yes, Herb had gone to the White House, but today I would have to fend for myself because I was counting on Herb and me having that final lunch.

The Parade

After that, Herb London introduced me to colleagues and associates at the Hudson Institute. I met Karl Rove — Bush's chief strategist, the architect of Republican electoral strategy. I met Douglas Holtz-Eakin, former Director of the Congressional Budget Office, the man who understood government economics better than almost anyone in Washington.

I met Armstrong Williams, conservative commentator, media personality, someone who could shape how ideas were presented to Black and conservative audiences simultaneously.

And more. Dozens more. Senior fellows at the Hudson Institute. Policy advisors. People with direct lines to Cabinet secretaries and White House senior staff.

Everyone wanted to know the same thing: Why is Herb London spending so much time with this Black guy? They'd see me in the hallways. In conference rooms. At more buffet tables. Always with Herb. Always deep in conversation. Always with documentation spread out between us.

Some were curious. Some were skeptical. Some were suspicious. But they all paid attention because Herb London doesn't waste time on people unless they have something worth hearing, and what I had was **Network** Law **& Order** — a comprehensive framework that solved three simultaneous problems:

Youth Unemployment and Aimlessness (which back then we just called "jobs for kids").

Critical Infrastructure Security (which back then was just "network security").

Economic Development That Reduces Debt (which back then was just "GDP growth through job creation").

I wasn't calling it "the new oil" yet. I didn't have the Rockefeller refinery metaphor. I didn't have the crude-to-refined framework..

I just had a really good jobs program that solved national security problems. And it was enough to convince Herb London in four lunch hours.

Hudson Institute

Chapter 9 reflects the author's personal recollections, professional interactions, and interpretations of events as he experienced them. Descriptions of conversations, meetings, and subsequent actions by others are presented as the author's perspective and understanding, not as assertions of endorsement, authorization, or official positions by any individual, institution, or government body.

CHAPTER 10

MARGE, WHAT MORE DO THEY WANT?

The Question

"Hey, Marge, why don't those people get together and do something for themselves? Hell, we've given them every damn thing. Some of them athletes walk around with diamonds around their necks and can't even read! Big fancy cars and no job! Always complaining, what more do they want? What about us?"

You've heard it. I've heard it. Many Black people in America have heard some version of this question — usually behind closed doors, sometimes in comment sections, occasionally to your face if the speaker feels emboldened enough.

The subtext is always the same: That "R" word again. Reparations. What more do they want?

It's a dismissive question designed to end the conversation, not start one. It assumes dependency. It ignores history. It pretends wealth gaps appeared by accident and can be wished away with bootstraps and personal responsibility speeches.

But here's what makes the question powerful: It contains a kernel of legitimate challenge. Why don't the wealthiest Black Americans pool resources and solve problems at scale?

That's actually a fair question. Not the racist framing around it. But the core question about collective action, pooled capital, and solving problems without asking permission from systems that have historically excluded us. That's the question I'm answering.

REPARATIONS

The JGI LLC Approach Under the United Nations Framework

The conversation about reparations for African Americans has long been polarizing — contentious in Congress, uncomfortable at dinner tables, and largely unresolved in the public square. Yet the international legal community has long moved beyond the debate of whether reparations are justified and toward the more practical question of what form they should take. The United Nations, through its Basic Principles and Guidelines on the Right to a Remedy and Reparation (A/RES/60/147), identifies five distinct forms of reparation for victims of gross human rights violations. The Jones Group International LLC (JGI LLC), through its **Network** Law **& Order** cybersecurity framework, situates itself within one specific category of that international taxonomy — one that offers what may be the most durable and broadly acceptable path forward.

The Five Forms of Reparations Recognized by the United Nations

The table below summarizes the five forms of reparation as defined under the UN Basic Principles, their legal meaning, and their potential application in the context of African American redress. The form selected by JGI LLC is highlighted.

Form	UN Definition	Application to African Americans
1. Restitution	Restore the victim to the original situation before the violation — restoration of liberty, employment, property, family life, and citizenship.	Return of confiscated lands, restoration of voting rights, and reversal of discriminatory policies that stripped economic and civic standing.
2. Compensation	Economically assessable damages: loss of earnings, property, opportunity, and moral damages incurred as a result of violations.	Direct monetary payment to descendants of enslaved persons for generational wealth denied by law through slavery and Jim Crow.
3. Rehabilitation	Medical and psychological care, legal and social services for those harmed by gross violations of human rights.	Investment in mental health, education, and community services in communities historically devastated by systemic racism.
4. Satisfaction & Guarantee of Non-Repetition ★ SELECTED	Public acknowledgment of violations, guarantees that violations will not recur, legislative and institutional reform, and measures that honor the dignity and truth of victims.	JGI LLC deploys NLO as a minority-owned innovation that enters the national digital infrastructure with legal protection against appropriation. No checks are written. Instead, a generation of youth cleans America's networks — using a Black inventor's framework.
5. Guarantees of Non-Repetition (standalone)	Institutional reform, civilian oversight, human rights training, and revision of laws that perpetuated violations to ensure they cannot recur.	Legislative reform preventing government seizure of minority-owned intellectual property; mandatory enterprise licensing as a structural safeguard.

"The Kind We Can All Live With"

Before unpacking the specific legal language of Satisfaction and the Guarantee of Non-Repetition, it is important to understand why JGI LLC selected this approach — and why it represents a form of reparations that transcends the familiar political divisions that have derailed every prior attempt at redress.

The debate over reparations has historically stalled over two parallel fears: Black Americans fearing that nothing meaningful will actually change, and White Americans fearing they will be asked to pay for something they personally did not do. Both fears are understandable. Both fears are also, in the JGI LLC model, addressed directly.

JGI LLC proposes no direct cash payments to individuals. Instead, it puts a minority-owned intellectual property framework — Panel of Experts-validated, technically rigorous, and commercially available — at the center of a national workforce development initiative called CyberCleaners. Young Americans between the ages of 16 and 22, across every demographic, will be trained to maintain and clean the nation's telecommunications networks using NLO's seven-role architecture.

Here is the key insight: Electrons do not play politics. A network node in Billings, Montana needs cybersecurity maintenance just as urgently as one in Baltimore, Maryland. A young Black teenager in Atlanta is not going to relocate to Wyoming to clean a network — but the network in Wyoming still needs cleaning. The beauty of a networked economy is that the work is distributed, the benefit is universal, and the tool doing the work belongs to a Black inventor. That is the kind of reparations we can all live with.

White families in rural communities benefit from network security maintained by a minority-owned framework. Black youth in urban centers gain access to a technology career pathway with a built-in sense of origin and pride. No community is asked to give up something they have. No community is asked to feel guilty. The ledger closes through contribution, not confiscation.

Satisfaction: The Word, The Meaning, The Mechanism

What "Satisfaction" Means in International Law

In the UN framework, Satisfaction refers to a category of reparation that is symbolic and structural rather than purely financial. It encompasses public acknowledgment of wrongdoing, verification of facts, official apologies, judicial and legislative rulings that restore the moral standing of victims, and measures that honor the truth and dignity of those who suffered. It is the form of reparation most directly aimed at healing the social fabric — not just compensating individuals, but changing the collective narrative.

What "Satisfaction" Means in the Context of NLO

In everyday American terms, Satisfaction means something simpler and more powerful: Both sides of a historical grievance come away from the table without feeling taken advantage of. That is a high bar. It is also, in the JGI LLC model, achievable.

Consider what Satisfaction means for the White workforce. When NLO is deployed inside a telecommunications company or a government network, no White employee loses a job that previously existed. The cybersecurity gaps that NLO fills were not being filled — or were being filled woefully, at enormous cost and with persistent vulnerability. NLO brings capability that was absent. The White engineer sitting in the next cubicle does not lose a seat at the table. The table gets larger.

Consider what Satisfaction means for the Black inventor. William B. Jones did not simply write a paper or draft a proposal. He spent decades at GTE Government Systems, earned electrical engineering degrees from Howard University and George Washington University, obtained FCC radio licenses as a teenager, served his country in uniform, and spent 10 years as a USPTO Patent Examiner before formally asserting his framework. When NLO is deployed, it is not deployed as a charity project or a set-aside program. It is deployed because it is the superior solution. That is a different kind of Satisfaction — the kind that carries pride.

Pride is not incidental to Satisfaction. It is central to it. When a minority inventor brings an innovation to a predominantly White industry and that innovation makes the industry function better, the moral arithmetic shifts. The narrative is no longer one of deficiency or dependency. It is one of contribution. Black America provided something the industry needed and could not produce on its own. That is Satisfaction in its fullest sense: a grievance addressed not through charity but through excellence.

The Guarantee of Non-Repetition: America's Long History of Taking

What "Guarantee of Non-Repetition" Means in International Law

The Guarantee of Non-Repetition is the forward-looking component of the Satisfaction category. It requires that the state take concrete steps — legislative, institutional, and structural — to ensure that the same violations cannot recur. In practice, this means reforming the laws, agencies, and practices that enabled the harm in the first place.

The Historical Pattern: America and Black Innovation

America has a well-documented and painful history of appropriating the inventions of Black Americans. From the refrigerator and the traffic signal to the blood bank and the cellular telephone, technology built upon foundational patents held by inventors whose names history forgot — the pattern is consistent. A Black inventor creates something of value. The apparatus of commerce, law, and government finds a way to absorb that value while leaving the inventor behind.

This is not an allegation. It is a documented pattern that includes formal mechanisms: patent interferences, discriminatory licensing practices, government seizure under national security authority, and corporate acquisition structures that erased inventor equity. NLO — Pentagon-validated, considered by six generals including former Chairman of the Joint Chiefs of Staff General Hugh Shelton and current National Security Advisor General Keith Kellogg — is not immune to this historical pattern simply because of its endorsements.

The Legislative Safeguard: Enterprise Licensing as Structural Protection

The Guarantee of Non-Repetition, as applied to JGI LLC and NLO, is therefore not merely symbolic. It must be structural. Legislation must be enacted that requires any government or commercial entity wishing to deploy NLO to do so through a formal enterprise license purchased from JGI LLC. This is not a novel legal concept. Enterprise licensing is a standard commercial practice. What is novel is making it a statutory requirement — specifically to prevent the government from doing what it has historically done to Black inventors: absorb the innovation into the federal apparatus and compensate the inventor with nothing but a footnote.

The legislative requirement serves three functions simultaneously. First, it creates a legal barrier to appropriation that cannot be circumvented by a procurement officer's discretion or an agency's internal reclassification. Second, it establishes a revenue stream for JGI LLC that is enforceable as a matter of law, not goodwill. Third, it sends a signal — publicly and permanently — that the United States government has acknowledged its pattern of appropriating minority innovation and has taken concrete legislative action to prevent its recurrence. That is a Guarantee of Non-Repetition in the fullest sense of the term.

The Affirmative Action Generation: Why This Matters Now

The individuals who carry the deepest psychological wounds from the failures of racial equity policy in America are not those who were enslaved — they are the generation immediately after. The first Black students admitted to previously White universities under affirmative action. The first Black employees promoted into white-collar positions in White-majority firms. The first Black officers integrated into previously segregated military units.

These individuals were not simply treated as equals entering a fair system. They were treated as experiments, tokens, and test cases. They were watched for evidence that their presence would degrade the institution. They were expected to represent their entire race in every meeting, every classroom, and every performance review.

The psychological cost of being "the first" was enormous but largely invisible to the institutions that benefited from their presence.

The Satisfaction model of reparations speaks directly to this generation's wound. Satisfaction says: Your contribution was real. The institution that doubted you needed what you brought. The innovation that followed your presence made things better — not worse — for everyone in the building. That acknowledgment, backed by legislative protection and commercial structure, is the form of reparations most commensurate with the harm experienced by the affirmative action pioneers.

It asks nothing of the White colleague except to acknowledge the obvious: the Table is better set because the inventor showed up.

Conclusion: A Reparations Model for the Digital Age

The **Network** Law & **Order** framework, deployed through JGI LLC, represents a form of reparations that fits the moment America is in — technologically urgent, economically productive, legally defensible, and morally coherent. It does not require White America to write a check. It does not require Black America to accept a symbolic gesture. It requires the government to buy what it already needs from the person who built it, and to pass a law ensuring it cannot steal it afterward.

That is Satisfaction. That is the Guarantee of Non-Repetition. That is the kind we can all live with.

The Answer

The Reparations Trust Group is a conceptual framework developed by the author. Seven leaders. The inventor of **Network Law & Order**, plus six of the wealthiest Black American billionaires. Their identities may be disclosed publicly in the future, but for now, understand the structure:

1. Total capitalization

The model envisions six participants each making multimillion dollar contributions. The inventor contributes intellectual property ownership — the NLO framework validated by the Pentagon, reviewed or evaluated by senior military officials, and not adopted within existing industry structures.

Seven leaders. One mission. This isn't another foundation. This isn't philanthropic "giving back." This is industrial capitalism at the scale that commands respect.

2. The Symbolic Power

When the Reparations Trust Group enters negotiations with governments, with the International Monetary Fund (IMF), with the World Bank, with Fortune 500 companies, we are not coming hat in hand. We are coming as peers.

That puts us at the same capitalization level as regional development banks. It's a hard-currency symbol that says: We're not here to ask. We're here to build.

3. Symbol of Unity in a Fragmented World

Until now, these six billionaires have existed on an island of social isolation from the 14 million African Americans they quietly long to uplift.

Their philanthropy has been filtered through foundations, boards, and non-Black intermediaries. Well-meaning but insulated staff have created distance between their wealth and the communities they most identify with. This "outsourced giving" has anesthetized impact — noble in intent, but disconnected in effect.

They've written checks. Foundations have executed programs. Reports have come back with metrics and testimonials. However, in many cases these wealthy African Americans can be three layers removed from the actual transformation.

This is different. **Network** Law **& Order** offers them direct participation in the very process that created their wealth: refining crude resources into value. Converting knowledge and systems into industrial capacity. Building infrastructure that generates recurring revenue while solving real problems. For the first time, they're not funding programs. They're building refineries.

4. Battling the IMF: Peer-Level Power

When governments need capital for infrastructure development, they go to the IMF or World Bank. They accept conditions. They cede sovereignty. They implement austerity measures that often harm the populations they're meant to serve.

The Reparations Trust Group offers an alternative. To countries struggling with youth unemployment, gang violence, cartel recruitment, and social instability: We'll train your young people as Cyber Cleaners. We'll secure your digital infrastructure. We'll create thousands of high-tech jobs that can't be outsourced. And you keep your sovereignty.

To governments concerned about cybersecurity vulnerabilities: We'll deploy **Network** Law **& Order** at scale. We'll customize it for your networks. We'll train your workforce. And we'll do it with capital that doesn't come with IMF conditions. This is economic warfare at the infrastructure level.

When a country can get capital, training, employment, and cybersecurity from the RTG instead of the IMF, they choose sovereignty. They choose development that benefits their people rather than foreign creditors.

What They're Actually Building: Retooling at Industrial Scale

Now that you understand the symbolic power and strategic positioning, let's talk about what the RTG actually builds.

Retooling: From Crude to Refined

In Chapter Two, I explained that crude oil was useless without refining infrastructure. Rockefeller became the richest man in history not because he found oil, but because he built the refineries that made crude oil valuable.

He controlled the infrastructure that transformed crude into gasoline, diesel, kerosene, and a hundred other products. Every barrel of crude that was refined generated profit. The more refineries he built, the more value he captured.

Network Law **& Order** (NLO) is crude oil. It exists. The model contemplates intellectual property licensing from JGI LLC to the

RTG. The Pentagon validated it. Six senior officials expressed interest in it. But in its current form, it's generic (crude) — a business process framework that works in principle but requires customization or refinement for deployment. And there's a second crude resource waiting to be refined: young people. Just like crude oil needs refining to become gasoline, diesel, and jet fuel:

Crude NLO needs refining to become:

Customized NCM for military networks.

Customized NCM for financial systems.

Customized NCM for healthcare networks.

Customized NCM for telecommunications carriers.

Customized NCM for AI data centers.

Customized NCM for government agencies.

Crude gamer skills need refining to become:

Focused Logistics.

Focused Switch.

Focused Transmission.

Focused Documentation.

Focused Security.

Focused IoTs (Internet of Things).

Focused Finance.

Rockefeller had one type of refinery. We're building two types simultaneously: NLO customization refineries and Cyber Cleaner training refineries. Both transform crude resources into refined products that generate recurring revenue.

Two Markets, One Solution

The genius of the refinery model: same infrastructure serves different markets with different positioning.

U.S. Market Positioning:

PRIMARY: Debt reduction through economic activity.

COLLATERAL: Youth aimlessness solved as a side benefit.

"This reduces national debt while securing networks."

International Market Positioning:

PRIMARY: Youth aimlessness and social stability.

COLLATERAL: Network security and economic development.

"This employs youth while securing digital infrastructure."

Same refinery infrastructure. Same NLO customization. Same Cyber Cleaner training. Different positioning for different markets.

This is industrial strategy at Rockefeller's level. Build once, sell everywhere, capture value from multiple revenue streams.

When bombs explode, we do a bomb damage assessment (BDA) for collateral damage. When NLO explodes into a network, we do a Collateral Advantage Assessment (CAA). The collateral advantages include: a reduction in youth aimlessness and a comprehensive plan to defend ourselves in cyberspace.

Reparations Through Network Law & Order

Network Law **& Order** (NLO) is the business process that provides reparations: the kind we can all live with. According to the United Nations, there are five types of reparations. For the purposes of this action, the Jones Group International Limited Liability Corporation, JGI LLC, adopts the UN category of Reparations classified as Satisfaction, and the Guarantee of Non-Repetition. This kind of reparations accounts for the descendants of slaves and the descendants of slaveholders.

Satisfaction: JGI LLC interprets this reparations attribute to mean "pride" and "fulfillment" in the apportionment. A minimum grumbling as to the unfairness or unjustness of the repair solution. An element of pride from the recipients of this type of reparations stems from the offering and introduction of technology from the African American community.

And: The "and" is a critical element of this kind of reparations because satisfaction must be sustained by some type of promise or guarantee.

Guarantee of Non-Repetition: The Department of Justice and the Small Business Administration were both created with clear founding purposes: protecting the rights of newly freed slaves and capitalizing African American small businesses. Over time, both agencies moved far from those original missions. For this reason, any guarantee must be written into law, protecting methods that create a preference-point system limited to the descendants of slaves and slaveholders.

This form of reparations is workable because a nation's future is built on innovation and technological growth, not cash handouts or job redistribution. Jobs are created by the very people the program is meant to help.

The comprehensive program to defend a country in cyberspace combats:

1. Youth Aimlessness: The Global Crisis

The author believes youth aimlessness is a significant contributing factor to domestic instability in America. It manifests differently around the world, but the pattern is identical:

America: Opioid addiction. Mass shootings. Gang violence. Domestic terrorism. Young people with no purpose, no trajectory, no ownership stake in society's success.

Europe: Radicalization. Lone wolf attacks. Youth joining extremist movements because extremism offers identity and purpose when society offers neither.

Latin America: Cartel recruitment. Gangs offer young people

structure, income, and belonging when the legitimate economy offers nothing.

Africa and Asia: Political instability. Militias recruit aimless youth. Civil unrest follows when generations have no economic path forward.

Every government on earth struggles with this. They fund "workforce development" programs that keep people busy without creating ownership. They offer training that leads nowhere. They provide jobs with no equity stake.

"The thing I miss the most from having all this happen is the internet. I mean, people don't realize what they have until it's gone. And, whew! Serious World of Warcraft *withdrawals, man, ha ha. Cause, in* World of Warcraft, *I'm awesome. I am a level 85 Paladin, huh, Tank and Healer. And in real life, I am a 14-year-old boy with nothing going for him."*

— "Poor Kids." PBS, November 20, 2012. Directed by Jezza Neumann. www.pbs.org/wgbh/frontline/film/poor-kids/

Level 85 Paladin. Tank and Healer. Master-level gaming skills that required hundreds of hours of strategic thinking, team coordination, resource management, and rapid decision-making under pressure.

"With nothing going for him." That's youth aimlessness in a white trailer in the Midwest. Same crude talent. Same need for refining. Same "nothing going for him" that leads some kids to opioids, some to gangs, and some — God help us — to walking into schools or churches or grocery stores with AR-15s.

The RTG refineries don't care about his race. They care about his gaming skills and his age. Can he map network topologies like he mapped World of Warcraft dungeons? Can he detect anomalies like he detected enemy attack patterns? Can he manage team resources like he managed raid groups?

If he can tank and heal in *World of Warcraft* at level 85, he can secure critical infrastructure in the real world. That's not charity. That's not a handout. That's recognizing crude talent when you see it and building the refinery infrastructure to transform it into refined professional capacity.

The RTG model works for him, too:

Age 14-16: Continue gaming, start pre-training modules online.

Age 16: Enter Cyber Cleaner training program.

Age 17-18: Deployed to first network security assignment at $65,000 per year.

From "nothing going for him" to a CyberCleaner in a global infrastructure. From that trailer to financial security. From aimless to purposeful.

And here's what matters politically: That kid lives in Middle America. His parents are supposed to believe in bootstraps and hard work and "no handouts."

But when the RTG opens a Cyber Cleaner training center in their region, and their son goes from "nothing going for him" to $65,000 per year, they don't care that it's funded by Black billionaires. They don't care that it's called Reparations. They care that their kid has a future.

This is how you build unstoppable national momentum. Black kids in Baltimore benefit. Latino kids in South Texas benefit. White kids in Appalachia benefit. Native American kids on reservations benefit. Any kid aged 16-22 with gaming skills benefits.

The refineries are color-blind in their hiring. The training is merit-based. The equity ownership is earned through work, not given through quotas.

But make no mistake: The RTG itself is not the reparations. African America's six wealthiest Black billionaires, plus the inventor of NLO, controlling a multi-million dollar capital fund. Black-owned refinery infrastructure. Black-led decision-making. Black wealth creation at an industrial scale.

The difference is that the infrastructure they build employs everyone's aimless youth. It secures everyone's networks. It reduces everyone's national debt burden. It prevents the domestic terrorism that happens when 14-year-old boys with "nothing going for them" find something to die for because they never found anything to live for.

That's the genius of retooling as reparations. It's not zero-sum. It's not taking from Peter to pay Paul. It's building refineries that create value for everyone while being owned and controlled by those who've been systematically excluded from building infrastructure at this scale.

When Marge's husband asks, "What more do they want?" She's probably thinking about her grandson in that trailer. She's probably worried about him, too. She probably sees the same aimlessness. The same "nothing going for him." The same fear that he'll end up another statistic.

The RTG is the answer to her grandson's problem, too. Not because they're asking her permission. Not because they need her approval. But because when you build refinery infrastructure at scale, the benefits ripple everywhere. That's not a bug. That's the feature.

2. The Answer for Marge

So when Archie asks, "What more do they want?" The answer is: "Nothing from you, America." The top six wealthiest Black billionaires in America just committed Capital to build a digital refinery infrastructure providing a realistic potential that may:

Reduce youth aimlessness.

Secure critical networks.

Reduce operating debt.

They're not asking for grants. They're not seeking approval. They're not waiting for permission. They're building refineries that transform crude talent into refined professional capacity while capturing value at every stage. They're doing it for themselves. At scale. With peer-level capital that puts them on equal footing with the IMF and World Bank.

And when historians write about the AI era, they won't write about another handout. They'll write about how seven Black leaders built the Rockefeller of the digital age — refinery infrastructure that everyone needed, but nobody else had the vision to build.

That "R" word isn't reparations-as-handout.

It's:

Reparations through Refineries.

Returns through Retooling.

Recognition as peers.

RTG building what governments couldn't.

And it started with seven leaders who got tired of hearing "What more do they want?" And decided to answer the question with refinery infrastructure. Not another program. Not another foundation. Not another philanthropic gesture. Simply, refineries. The kind Rockefeller built.

Chapter 10 presents the author's opinions, strategic proposals, and forward-looking concepts. References to capital commitments, projected outcomes, or institutional interactions reflect aspirational planning and the author's perspective, not binding commitments or guarantees, nor allegations of misconduct by any individual, organization, or government body.

CHAPTER 11

JEKYLL 2.0

The Herb Can Do It Symposium

Jekyll 1.0. November 1910. Jekyll Island, Georgia.

Seven men met in secret. Senator Nelson Aldrich, A. Piatt Andrew, Frank Vanderlip, Henry Davison, Charles Norton, Benjamin Strong, and Paul Warburg. They represented one quarter of the world's wealth. They didn't ask the president's permission. They designed the Federal Reserve System and presented it as infrastructure the country needed. The government adopted it because the alternative was chaos.

History does not remember them as men who asked. History remembers them as men who built.

Jekyll 2.0. 2026. Jekyll Island, Georgia.

The Reparations Trust Group does not meet to ask. We meet to build.

The objective is singular: Retool **Network** Law & **Order** into the refinery infrastructure that secures critical networks, eliminates youth aimlessness, and generates returns that make the argument for themselves. Six four-hour sessions over four days. The structure mirrors Chapter Nine's "Herb London Can Do It" symposium, now expanded to match the scale of what we are actually building.

THE GATEWAY QUESTION

Every major infrastructure deployment in American history has had a prototype state. The transcontinental railroad had California. The TVA had Tennessee. The interstate highway system had Kansas. The state that hosts the first NLO refinery doesn't just get a facility. It gets the template. It gets the patent position. It gets the workforce pipeline. It gets the lease revenue. And it gets the permanent designation as the birthplace of the infrastructure that secured America's digital future.

Maryland is the logical choice. Proximity to the Pentagon, NSA, and the federal contracting apparatus that will accelerate NLO deployment is not incidental — it is the entire point. Maryland already hosts more cleared defense contractors per square mile than any state in the union. The NLO refinery would be a logical next step for Maryland. It is the natural next step in it. The question is not whether Maryland wants this. The question is whether Maryland leadership moves before another state does.

THE NATIONAL SECURITY IMPERATIVE

The Telecommunications Industry Association represents the backbone of American critical infrastructure. Its member companies are designated National Infrastructure Protection Plan assets - NIPP classified — meaning the federal government has already determined they are essential to national security. That designation carries weight. It also carries responsibility.

An association whose members are national security assets must itself operate as a national security asset. That means NLO is not optional for TIA. NLO is the operational standard that makes the designation real. Member companies will demand a 30 percent reduction in operational costs. They will demand the lease revenue that comes from hosting CyberCleaner digital classrooms. They will demand the customer retention that comes from offering the only network configuration management framework validated by a DoD Panel of Experts.

TIA can facilitate that deployment or watch its members negotiate directly. Either way, NLO deploys.

THE GATEWAY REFINERIES: PANAMA AND AFRICA

The United States is not the only country whose critical infrastructure is vulnerable. It is simply the first country with a validated solution.

Panama controls the corridor between the Atlantic and Pacific. A Caribbean gateway refinery positions NLO at the intersection of hemispheric telecommunications traffic. African nations representing 54 countries and 1.4 billion people are building digital infrastructure from the ground up. They are not locked into legacy systems. They are not burdened by fragmented standards left over from the AT&T breakup. They are ready for NLO in ways that Western nations, tangled in their own bureaucratic resistance, are not.

The embassies that understand this earliest will have the clearest path to a Jekyll 2.0 invitation. The window is not permanent.

THE REPARATIONS TRUST GROUP

The Reparation Trust Group consists of the wealthiest African American billionaires making significant capital investment in the refinery infrastructure that everyone will need and no one else has the vision to build.

The RTG understands infrastructure investment, they understand the Rockefeller model. The Rockefeller model was never about oil. It was about controlling the refinery. The man who controls the refinery controls the market. NLO is the refinery. The network is the new oil.

The RTG chair is not a ceremonial position. The chair of the RTG is the chair of the entity that deploys the Federal Reserve of digital infrastructure. That seat at Jekyll 2.0 is not simply an invitation. It is a position in history.

INVITEES

Jekyll 2.0 convenes the Reparations Trust Group, Maryland leadership, international gateway representatives, and Keith Kellogg as representative of the United States Government. Maryland leadership organizes the summit — not to request funding, but to position Maryland as the U.S. prototype state before another state does.

EPILOGUE

LET IT BE

There was a moment when I stopped wondering if racism was embedded in cybersecurity and started knowing it.

The Chief Technology Officer for the District of Columbia looked across the table and said, without hesitation, without shame, and without whispering: "I can't believe that these jungle bunnies would even know anything about CM. Hell, I barely know what you're talking about."

I stared. My Vice President's son stared at his father. His father stared at me.

Three people in that room heard it. No one said a word.

On the way back from the briefing, my Vice President - who is white - and his son and I looked at each other and said what we were all thinking. The CIO had said it like it was nothing. No hesitation. Noshame. The Chief Technology Officer of the District of Columbia had just called D.C. constituents jungle bunnies, and it didn't even slow the CIO down.

I still remember that moment as clearly as the day it happened.

This was not bureaucratic resistance. Not institutional inertia. Not a funding problem or a timing problem or a political problem.

This was the problem.

The title of this book was born in that room.

Four presidents had their chance.

Bush heard it through Herb London. Obama heard it through ten contacts and then watched a National Security Advisor appropriate my company name. The author contacted the Secretaries of Commerce of three successive administrations. None responded.

Twenty-five years. Four administrations. Zero results.

Not because they were bad people. Because the system was designed to prevent someone like me from building infrastructure that bypasses traditional gatekeepers and creates wealth directly for workers.

So I stopped asking.

The Reparations Trust Group needs no presidential approval. Seven leaders. Significant capital. Refinery infrastructure that solves youth aimlessness while securing critical networks and generating returns that make the argument for us.

Let it be.

After six generals validated the NLO claimed benefits. After Herb London carried it to the Bush White House. After four administrations looked away. After ten years of examining inadequate cybersecurity patents while the solution sat in my files.

Let it be.

We build the refineries. Countries compete for access. The White House becomes a customer.

That is not surrender. That is the only lesson worth learning from nineteen years of watching the system protect itself.

When historians write about the AI era, they will not write about another handout. They will write about how seven Black leaders built the Rockefeller of the digital age - refinery infrastructure that everyone needed, and only one man had spent 25 years proving could work.

THE END

ACKNOWLEDGEMENTS

THE NETWORK LAW & ORDER HALL OF FAME

Network Law & **Order** did not come into existence because institutions approved it.

It survived because individuals — sometimes quietly, sometimes courageously — chose to stand beside an idea before it was safe to do so.

This Hall of Fame recognizes those who acted when acting carried consequences.

White Americans Who Chose Character Over Convenience

1965 - Bill Belch — Owner, WIDU Radio

When I was thirteen years old, Bill Belch made me a promise: If I obtained a Third Class FCC license, he would give me a teenage radio show. I earned the license. He kept his word. That lesson — merit honored by integrity — never left me.

1969 - Dr. Wesley Wallace — Dean, Department of Radio and Television, UNC Chapel Hill

Dr. Wallace traveled to Fayetteville, North Carolina, to attend my high school graduation and family dinner. Long before titles or inventions, he validated potential.

Rest in Peace.

1985 - Ken Davidson — Defense Information Systems Agency (DISA)

A direct witness to the birth of **Network** Law & **Order** and Network Configuration Management. Ken recognized the significance of what was being created when few others did. A witness to the events documented in *The Racism of Cybersecurity*.

1998 - Cliff Shoemaker and Family

My lawyer, financial advisor, defender, and trusted witness, Cliff stood beside me through battles most professionals would have avoided. He remains the only White man I personally know suffering from PTSD connected to witnessing the realities described in TROC.

2000 - Ken Wilkinson — Investor and Friend

An early investor who became a genuine friend. A committed conservative who believed both in **Network** Law & **Order** and in its creator. We argued hard, disagreed often, and respected each other completely.

Rest in Peace.

1999 - Cord Sterling — Congressional Aide

The first legislative staff supporter willing to engage **Network** Law & **Order** seriously. He helped drive the Panel of Experts effort and opened doors normally inaccessible to outsiders. I received what amounted to a million dollars' worth of access — all in exchange for Union Station parking.

2000 - Jim Wilkins — Federal Aviation Administration

An unexpected angel investor who, for two years, encouraged me daily to stay focused on the work and ignore the turbulence surrounding it. His steady belief mattered more than he likely knew.

Rest in Peace.

2006 - Dr. Herb London — Hudson Institute

He tried — and he kept his promise. Dr. London carried **Network** Law & **Order** to the White House, becoming one of the very few willing to test the idea at the highest levels of national discussion. A genuine witness to this history.

Rest in Peace.

Black Americans Who Built the Foundation

Dr. Vanilla K. Wilson — My Mother

She established the governing equation of my life:

If you can see and believe, you can achieve and receive. She believes in me.

1968 - Atlanta Field Office, NAACP Atlanta Georgia.

When a sixteen-year-old student boarded at Elkins Institute pursuing a First Class FCC license, the NAACP intervened to ensure accommodation and fairness. That intervention helped launch the technical journey that ultimately led to **Network** Law & **Order**.

Rest in Peace to those who carried that struggle.

1978 - Dr. Jean K. Sinkford — Dean, Howard University Dental School

She insisted I pursue engineering when my path was uncertain. That decision changed my life — and indirectly shaped this entire body of work.

1981 - Dr. Tepper Gilll — Howard University CAR Program

The educator who taught me how to truly understand mathematics. Without that foundation, **Network** Law & **Order** would never have been possible.

2000 - Shirley Fields — Chief Information Officer, Defense Information Systems Agency (DISA)

As DISA's Chief Information Officer, Shirley Fields placed **Network** Law & **Order** before the agency no fewer than three formal times, ensuring the concept received institutional visibility and serious consideration. She stands as a direct witness to the events documented in *The Racism of Cybersecurity*.

1999 - Gladwyn Goins — Former Director, California Securities and Exchange Commission

After twenty-five years, I finally accepted and acted upon his advice. Some mentors plant seeds whose meaning becomes clear only decades later.

1986 - Don Lewis — GTE Government Systems

He said only two words, "I'll help." And he did. A witness to the realities recorded in this book.

1985 - George Luckett — Defense Information Systems Agency

Known simply and proudly by his signature: "DISA LUCKETT."

A civil servant who loved his country far more than his country loved him. His operational thinking and quiet professionalism helped shape what would become **Network** Law & **Order**. In many respects, he stands as the true grandfather of NLO.

Rest in Peace.

Author

William Benjamin Jones

"The darker races of mankind, and the black race in particular, will keep the white race busy for the next hundred years, in defending the interests of white supremacy. The black singer is coming with his song, the poet with his dreams, the sculptor with a conception of beauty and awe, and the scholar with his truth. The greatest marathon race of the ages is about to begin between the white race, and the darker races of mankind. What Jack Johnson seeks to do to Jim Jeffries, in the roped arena, will be the ambition of Negroes of every domain of human endeavor".

The Reverend Reverdy Ranson, 1910

The Business Process:

Network Law & Order™

Next Generation

Network Configuration Management

www.ingramcontent.com/pod-product-compliance
Ingram Content Group UK Ltd.
Pitfield, Milton Keynes, MK11 3LW, UK
UKRC032148290726
14090UKWH00012B/501